AF541594

Towards Sustainable Agriculture
Policies and Practices

NIPA® GENX ELECTRONIC RESOURCES & SOLUTIONS P. LTD.
New Delhi-110 034

About the Editor

Suresh Chandra Patnaik is an eminent agriculturist with a postgraduate degree in agriculture, who has made significant contributions to the field throughout his illustrious career. He dedicated his early years (1968-1978) to rice research, teaching, and extension at Odisha University of Agriculture and Technology (OUAT). Transitioning to government service from 1978 to 2003, he played a pivotal role in enhancing the capabilities of farmers and extension workers across Odisha's agricultural districts. His efforts in policy formulation, budgeting, and planning significantly influenced the production and seed distribution systems within the state. From 2003 to 2015, Patnaik served as a consultant for UNDP-GOI project focusing on natural resource management and worked with the IFAD-DFID-assisted Odisha Tribal Empowerment and Livelihoods Program (OTELP). Since 2015, he has been a consultant for RICOR, specializing in agriculture and decentralized planning, with a commitment to uplifting marginalized rural and urban communities. He has been a board member of the Tagore Society of Rural Development-Odisha Projects since 2013, where he has continued his efforts to support these communities.

Patnaik began his journey with IFAD in 2010 by participating in a review mission for agricultural development under the Jharkhand Tribal Development Project (JTDP). Over the years, he held various roles, including serving as the State Implementing Officer for the IFAD-ICRISAT projects under OTELP from 2007 to 2013, where he oversaw multiple initiatives focusing on legume crops and micro-irrigation. As a consultant for RICOR, he contributed to the development of a decentralized planning manual for the Odisha government, aiding district planning efforts with support from UNICEF. In 2016, he transitioned to the Administrative Staff College of India, where he continued to work as a consultant, conducting Training Needs Assessments (TNAs) and training stakeholders in district planning methodologies.

From 2013 to 2019, Patnaik served as a member of the Board of Management at OUAT, approved by the Agriculture and Farmers' Empowerment Department of the Government of Odisha. Due to his valuable contributions, he was re-nominated to the OUAT Board of Management in October 2019 for a second term. His efforts have been instrumental in shaping OUAT into a modern agricultural university in eastern India.In May 2020, he authored a concept note on skill development for the Odisha Mineral Bearing Areas Development Corporation (OMBADC), which received positive feedback from the Government of Odisha. This resulted in an allocation of Rs 146.92 crore in funding by OMBADC for OUAT, aimed at supporting one-year certificate skill training courses for farmers and entrepreneurs in agriculture and allied sectors.

Throughout his career, Patnaik has demonstrated a commitment to continuous learning and professional growth, actively seeking opportunities to expand his expertise. He has participated in various training programs and seminars, including visits to successful Sloping Agriculture Land Technology (SALT) models in the Philippines and discussions with scientists at the International Rice Research Institute. He has engaged internationally, attending workshops in Vietnam, Nepal, and China, where he contributed to discussions and published research papers. In India, he has participated in training programs organized by esteemed institutions such as ICRISAT and ICAR, focusing on groundnut and rice research. His collaborative efforts with organizations across India, including BAIF, Karnataka Bhaktivedanta Academy, and MYRYDA, showcase his dedication to knowledge exchange and skill-building.

Patnaik's literary contributions to agriculture include his debut book, "Promotion of Sustainable Agriculture and Horticulture Development," published in 2009, and his second publication, "My Dream OUAT: Transformation towards a Modern Agricultural University in India," released in 2019. He has also contributed to six other notable works: "Trainer's Training Manual on Agriculture" by OTELP (2010); "Trainer's Training Manual on Horticulture (Part I)" by OTELP (2010); "Trainer's Training Manual on Horticulture (Part II)" by OTELP (2010); "Manual on Land and Water Management" by OTELP (2009); "Successful Interventions of RKVY in OTELP Areas of Odisha" (2014); "Horticulture for Rural Development" by Dr. Sanghamitra Patnaik (2019). Additionally, he crafted a comprehensive brochure titled "OUAT - The New Horizon: Facilitation to Regulate Policy Matters and Feedback on Ground Reality of OUAT," covering the period from October 2019 to June 2021. Most recently, he contributed to the book "Agriculture Policy and Development in Eastern India," published in 2023.

Suresh Chandra Patnaik's extensive career reflects his unwavering dedication to agricultural development, policy formulation, and community upliftment, making him a significant figure in the advancement of agriculture in India.

Dt. 04.06.2024 **Suresh Chandra Patnaik**

Towards Sustainable Agriculture
Policies and Practices

Suresh Chandra Patnaik

NIPA® GENX ELECTRONIC RESOURCES & SOLUTIONS P. LTD.
New Delhi-110 034

**NIPA® GENX ELECTRONIC
RESOURCES & SOLUTIONS P. LTD.**

101,103, Vikas Surya Plaza, CU Block
L.S.C. Market, Pitam Pura, New Delhi-110 034
Ph : +91-11-43860225, Mob.: +91 9717133558, 9540816132
E-mail: newindiapublishingagency@gmail.com
Website: www.nipaersources.com

Print ISBN: 978-93-58874-71-6
ebook ISBN: 978-93-58874-28-0

NIPA® also publishes books in a variety of electronic formats. Some content that appears in print may not be available in electronic books, and vice versa.

Composed and Designed by NIPA®.

ଡଃ. ଅରବିଂଦ କୁମାର ପାଢ଼ୀ
Dr. Arabinda Kumar Padhee, IAS
Principal Secretary

Foreword

Agriculture remains the cornerstone of India's economy, engaging nearly half of its workforce. Despite its significant GDP contributions, the sector faces a myriad of challenges, including fragmented land holdings, unpredictable weather patterns due to climate change, and the need for sustainable practices. In this region, rich in natural resources and diverse agro-climates, there is immense potential but also specific obstacles like recurring natural disasters, inadequate infrastructure, and socio-economic disparities. At this critical juncture, agriculture in this area must navigate the balance between tradition and modernization.

Throughout India, efforts to enhance agricultural productivity and address climate change encounter obstacles in fully tapping into the sector's potential for poverty reduction and livelihood improvement. Nevertheless, sectors like fisheries, aquaculture, agroforestry, livestock rearing, and poultry farming have emerged as crucial pillars of the agricultural economy, offering avenues for diversification, income, generation, and food security.

Against this backdrop, Sri Patnaik's book **"Towards Sustainable Agriculture: Policies and Practices"** not only delves into conventional agriculture but also covers allied sectors and existing policies. Its aim is to shed light on the progress, challenges, and evolving dynamic role of financial institutions in the eastern region's development, as well as land tenure issues and their impact on agricultural productivity and the national agroforestry policy.

Authored by seasoned experts, this edition bridges these and insights for educators, scholars, students, extension officers, policymakers, and the broader community committed to advancing agricultural progress in the region. Sri Suresh Chandra Patnaik, expert in agricultural extension and development and a distinguished member of the Board of Management at Odisha University of Agriculture & Technology (OUAT), Bhubaneswar, has spearheaded this commendable effort, emphasizing the collaborative approach needed to drive agricultural sustainability and development forward.

(Dr. Arabinda Kumar Padhee)

Bhubaneswar
Dated: 4th June, 2024

Department of Agriculture and Farmers' Empowerment
Krushi Bhavan, Bhubaneswar, Odisha, India, Pin-751001, Phone : 0674-2914411 (155333), e-mail: agrsec.or@nic.in

Preface

India's agricultural development since Independence can be broadly divided into following phases, each characterized by different policies, strategies, and outcomes. Here's an overview: Pre-Green Revolution Era (1950-1960s): Agrarian economy dominated by the traditional farming practices. Focus was on land reforms for equitable distribution of land. Community development programs were initiated with the aim of enhancing agricultural practices. The establishment of agricultural universities and research institutions were made to further advance the field. Green Revolution (1960s-1980s): Modern agricultural techniques, alongside the adoption of high-yielding seed varieties, chemical fertilizers, and pesticides, have led to a substantial boost in crop productivity, particularly evident in wheat and rice cultivation. Irrigation facilities were expanded. These led to self-sufficiency in food grains and transformed India from a food deficit to a food surplus nation. Post-Green Revolution (1980s-1990s): Focus was made on diversification of agriculture beyond wheat and rice. Efforts were made to boost horticulture, dairy farming, poultry, and fisheries. The inception of technology dissemination initiatives such as Krishi Vigyan Kendras (KVKs) occurred. Government initiated improvement in rural infrastructure and market access. Liberalization and Globalization (1990s onwards): Economic reforms were made which led to liberalization of agriculture, allowing for greater private sector participation. State support for agriculture was withdrawn, subsidies were reduced, and agricultural price control mechanisms were dismantled. Export-oriented agriculture was prioritized, alongside the promotion of contract farming. Challenges were on farmer distress, agrarian crisis, and widening rural-urban disparities. Sustainable Agriculture (2000s onwards): Acknowledged the environmental issues and embraced sustainable agricultural methods. Organic farming, conservation agriculture, and integrated pest management were promoted. Initiatives were made to enhance soil health, water conservation, and biodiversity conservation. Emphasis was done on climate-resilient agriculture in the face of climate change challenges. Recent Developments: The integration of technology-driven solutions like precision agriculture, remote sensing, and GIS mapping was implemented. Schemes like Pradhan Mantri Krishi Sinchayee Yojana (PMKSY) for irrigation and Pradhan Mantri Fasal Bima Yojana (PMFBY) for crop insurance were introduced. Focus was made on doubling farmers' income through various measures including value addition, market linkages, and improved productivity. Across these stages, persistent challenges have included land fragmentation, inadequate infrastructure, market fluctuations, and agrarian distress.

The livelihoods of a significant portion of India's population are supported by agriculture and allied sectors, a role that may remain dominant even in the future. The sector proved a silver lining in the pandemic period registering a positive growth in the covid times. India registered agriculture growth of 3.6 percent during 2020-21 and 3.9% in 2021-22 as per economic survey of India, 2020-21. Yet it faces various structural challenges to be addressed to make it profitable. Recognizing the interconnectedness of various allied sectors with agriculture, it becomes imperative to adopt a holistic approach towards policy formulation and implementation. In this book, we traverse the historical trajectory of agricultural and allied sector development, delineating the phases of evolution and assessing the current status quo. From the flourishing fisheries and aquaculture development to the pivotal role of livestock in bolstering the income of small and marginal farmers, each chapter offers profound insights into the challenges and opportunities inherent in these domains. Moreover, we place a special emphasis on gender inclusivity and empowerment, acknowledging the indispensable contribution of women in agriculture. Through a nuanced analysis of their roles and the strategies for gender mainstreaming, we strive to engender a more equitable and sustainable agricultural paradigm. Furthermore, the book meticulously examines the tools and approaches employed in agricultural extension services, elucidating the challenges confronting the sector and the role of financial institutions in catalysing its growth. National Agroforestry Policy 2014 in India which aims to enhance agricultural livelihoods by integrating agriculture, livestock farming, and forestry activities & mitigating climate change through carbon sequestration have also been covered in the book. Land tenure issues and their implications for agricultural productivity are also scrutinized, underscoring the imperative of land reforms in fostering rural prosperity.

As we embark on this scholarly journey, it is my fervent hope that the book entitled "**Towards Sustainable Agriculture: Policies and Practices**" serves as a beacon of knowledge, guiding policymakers, practitioners, and academicians towards informed decision-making and transformative action. I extend my heartfelt gratitude to all the contributors whose expertise and dedication have enriched this volume, and I am confident that their collective wisdom will inspire meaningful change in the agricultural landscape of our nation.

Editor

Acknowledgement

On September 25, 2020, I had the privilege of delivering a talk on Agriculture Development in Odisha, focusing on marketing, during a webinar hosted by the esteemed organization, Think Tank, Bhubaneswar. Dr. Saurav Garg, IAS, Former Principal Secretary of the Agriculture and Farmers' Empowerment Department, Government of Odisha, delivered the keynote address. I extend my heartfelt thanks once again to Think Tank for the opportunity.

Following this event, I contemplated publishing a book on various aspects of agriculture in Odisha. However, after discussions with Dr. Ashok Kumar Sahoo, Former Professor, Department of NRM, College of Forestry, OUAT, it was decided to publish a comprehensive book on "Agriculture Policy and Development in Eastern India." This book was published with an ISBN by NIPA, New Delhi, in 2023. On the day of its release, several invitees suggested publishing a book to cover allied sectors of agriculture, as the current edition did not address them, particularly any policy development in India. This decision was made to further engage stakeholders in enhancing the livelihood of farmers with reference to agricultural developments.

The book titled "**Towards Sustainable Agriculture: Policies and Practices**" is also the result of collaborative efforts from several authors, whose contributions at various stages have been invaluable. I extend my appreciation to these scholarly authors for their contributions to different chapters of this book.

I extend my deepest appreciation to Professor Ashok Kumar Sahoo, former Professor, OUAT, and Dr. Amiya Kumar Behera, Chairman, RICOR, Bhubaneswar, for their unwavering support, invaluable insights, and meticulous guidance throughout the creation of this book. Their commitment to excellence has been evident in our countless discussions and their insightful suggestions. Their expertise and dedication have truly enriched this work, and I am profoundly grateful for their enduring collaboration.

Bhubaneswar
Dt. 04.06.2024

Suresh Chandra Patnaik

ICAR: Indian Council of Agricultural Research)

ICAR-CIFA: Indian Council of Agriculture Research- Central Institute of Freshwater Aquaculture

ICAR-CIFRI: Indian Council of Agriculture Research-Central Inland Fisheries Research Institute

ICAR-CMFRI: Indian Council of Agriculture Research-Central Marine Fisheries Research Institute

ICAR-CPCRI: Indian Council of Agriculture Research- Central Plantation Crops Research Institute

ICAR-NBAGR: Indian Council of Agriculture Research - National Bureau of Animal Genetic Resources

ICAR-NRRI: Indian Council of Agriculture Research Institute- National Rice Research Institute

ICT: Information Communication Technology

IFFCO: Indian Farmers Fertilizer Cooperative Limited

IFS: Integrated Farming System

IIPM: Indian Institute of Plantation Management

IJCSBE: International Journal of Case Studies in Business, IT, and Education

IMC: Indian Major Carps

IRRI: International Rice Research Institute

JICA: Japan International Cooperation Agency

KALIA: Krushak Assistance for Livelihood and Income Augmentation

KCC: Kissan Credit Card

Kg/ha/yr: Kilogram per hectare per year

Kg: Kilogram

KSHAMTA: Knowledge Systems and Homestead Agriculture Management in Tribal Areas.

KVK: Krishi Vigyan Kendra

LAMPS: Large Area Multi-Purpose Societies

LHRH: Luteinising Hormone-Releasing Hormone

MANAGE: National Institute of Agricultural Extension Management

MELA: Madras Estates Land Act

MFIs: Micro Finance Institutions

MGNREGA: Mahatma Gandhi National Rural Employment Guarantee Act

MIDH- Mission for Integrated Development of Horticulture

MKSP: Mahila Kisan SashaktikaranPariyojana

MKUY: Mukhyamantri Krushi Udyog Yojana

MKUY: Mukhyamantri Krushi UdyogYojana

MLP: Major Livestock Products

mm: Millimetre

MOVCD: Mission on Organic Value Chain Development

MPEDA: Marine Products Export Development Authority
NAAS: National Academy of Agricultural Sciences
NABARD: National Bank for Agriculture and Rural Development
NADCP: National Animal Disease Control Programme
NADRS: National Animal Disease Reporting System
NAPCC: National Action Plan on Climate Change
NARES: National Agriculture Research and Education System
NARI: Nutri-sensitive Agricultural Resources and Innovation
NARS: National Agriculture Research System
ND: Newcastle Disease
NDP: National Demonstration Project
NEP: New Education Policy
NFDB: National Fisheries Development Board
NFSM: National Food Security Mission
NGOs: Non-Government Organizations
NHB: National Housing Bank
NHM: National Horticulture Mission
NICRA: National Innovation on Climate Resilient Agriculture
NICRA: National Innovation on Climate Resilient Agriculture
NIPHM: National Institute of Plant Health Management
NLM: national Livestock Mission
NMAET: National Mission on Agricultural Extension and Technology
NMOOP: National Mission on Oilseeds and Oil Palm
NMSA: National Mission on Sustainable Agriculture
NPCBB: National Project for Cattle & Buffalo Breeding
NPOP Certification: National Programme for Organic Production Certification
NRAA: National Rainfed Area Authority
NRCAF: National Research Centre for Agroforestry
NRLM: National Rural Livelihoods Mission
NSDP: Net State Domestic Product
NSO: National Sample Survey
NSPAAD: National Surveillance Programme on Aquatic Animal Diseases
OD: Odisha
OLR Act: Odisha Land reform act
OLRDS: Odisha Livestock Resource Development Society
OMFED: Odisha Cooperative Milk Producers Federation Ltd
OMM: Odisha Millet Mission
OPOLFED: Odisha State Poultry Producers Co-operative Marketing Federation Ltd
OUAT: Odisha University of Agriculture & Technology

PACs: Primary Agricultural Credit Societies
PDS: Public Distribution System
PED: Professional Efficiency Development
PGS Certification: Participatory Guarantee System Certification
PKVY: Paramparagata Krishi Vikas Yojana
PMKSY: Pradhan Mantri Krishi Shinchayee Yojana
PMKVY: Pradhan Mantri Kaushal Vikas Yojana
PMS: Pregnant Mare Serum
PNB: Punjab National Bank
PPP: Promoting Public Private Partnership
PPR Disease: Peste des Petits Ruminants Disease
PSSB: Professional Standards Setting Body
PVCF: Poultry Venture Capital Fund
RAD: Rainfed Area Development
RBI: Reserve bank of India
RGCA: Rajiv Gandhi Centre for Aquaculture
RIDF: Rural Infrastructure Development Fund
RKVY: Rastriya Krishi Vikas Yojana
RKVY-RAFTAAR: Rashtriya Krishi Vikas Yojana-Remunerative Approaches for Agriculture and Allied Sector Rejuvenation
RPL: Recognition of Prior Learn
RRBs: Regional Rural Banks
RRC: Regional Research Centre
SAMETI: State Agricultural Management & Extension Training Institute
SAP: Sustainable Agricultural Production System
SAU: State Agricultural University
SCB: State Co-operative Bank
SCSP: Scheduled Caste Sub Plan
SDG: Sustainable Development Goal
SFDCL: State Fisheries Development Corporation Ltd.
SFP: State Focus Paper
SHGs: Self-Help Groups
SMAM: Sub-Mission on Agricultural Mechanization
SMSP: Sub Mission on Seed and Planting Material
SPCA: State Society for Prevention of Cruelty to Animals
T & V Programme: Training and Visit Programme
t: Tonne
TKS: Tata Kisan Sansars
TKVK: Tata Krishi Vikas Kendras

TSP: Tribal Sub-Plan

TTK: Tata Kisan Kendra

UEBP: University Extension Block Programme

UGS: Utkal Gomangal Samiti

UP: Uttar Pradesh

VATICA: Value Addition and Technology Incubation Centres in Agriculture

WB: West Bengal

WBFCL: West Bengal Fisheries Corporation Ltd.

WBI: Wood Based Industries

WFP: World Food Programme

WTO: World Trade Organization

1

Phases of Development and Current Status of Agriculture and Allied Sectors in India and Eastern States

Suresh Chandra Patnaik

Former Agriculturist and Managing Director, OSSC, Agriculture and Farmers' Empowerment Department, Government of Odisha
Member, Board of Management, Odisha University of Agriculture & Technology

1. Introduction

The agricultural sector plays a pivotal role in driving economic development in India, a nation with a population of 1.39 billion, making it the world's second most populous country according to the United Nations' World Population Prospects 2019. With 54.6 percent of the workforce engaged in agriculture as per the 2011 Census, and contributing 18.6 percent to the Gross Value Added (GVA) in 2021-22, its significance cannot be overstated. Notably, women play a substantial role in this sector, with rural females exhibiting a higher workforce participation rate of 41.8 percent compared to 35.31 percent for urban women (MoSPI, 2017). In rural communities, agriculture and allied sector is the primary source of livelihood that includes 80 percent of all economically active women, out of which 33 percent constitute agricultural labour force and 48 percent are self-employed farmers. Rural women are engaged at all levels of the agricultural value chain; i.e., production- pre-harvest, post-harvest, processing, packaging, marketing to increase productivity in agriculture (Nilam Patel etal, 2022). As per Pingali *et al.* (2019), the ratio of women to men working in the agricultural sector has increased over the time and made a greater amount of contribution to GDP per capita. They are the momentous demographic group for a sustainable food system (FAO, 2011). Empowering women in agriculture through reforms aimed at ensuring equal access to resources, skill development, and opportunities is projected to increase agricultural output in developing countries by 2.5 to 4 percent (FAO, 2011). However, there remains a significant income disparity between agricultural and non-agricultural workers, with poverty and undernutrition disproportionately affecting agricultural labourers and small-scale farmers, leading to rural distress (Chand *et al.*, 2015; Chand, 2019). Over the past 75 years, India has undergone a remarkable transformation from being food-scarce to becoming a net exporter of food, largely due to the implementation of science-backed agricultural revolutions by the government. Notably, initiatives like the Green Revolution have significantly contributed to this progress. Despite a staggering 237 percent increase in the population, the country has managed to more than double its per capita food production over the last 50 years, as noted by Ramesh Chand in his report on Agricultural Challenges and Policies for the 21st Century (NABARD, 2022). However, looking ahead, India faces the daunting task of meeting the food demands of an expected population exceeding 1.6 billion by 2050. The Indian Council of Agricultural Research (ICAR, 2022) highlights the necessity of achieving a minimum 4% annual growth in agriculture to address this challenge. Yet, several hurdles persist across various sectors of Indian agriculture, including issues related to growth, sustainability, efficiency, and equity. In addition to these challenges, ensuring food and nutritional security, tackling climate change adaptation and mitigation, enhancing livelihoods, employment opportunities, and elevating the standard of living for agricultural communities are pressing concerns. While the government is actively engaged in addressing these issues through ongoing programs and policies, there's a recognized need for long-term strategies that focus on enhancing farmers' income, ensuring nutritional security, promoting sustainable food systems, and realigning agricultural policies accordingly.Given India's diverse agroecosystems, characterized by varying climatic, soil, and natural features, a national-level dialogue that encompasses diverse perspectives is crucial for devising suitable solutions for the future of agriculture in the country.

Agriculture policies & phases of development in India: Prior to the announcement of the first ever National Agriculture Policy in July, 2000 by the Government of India, agriculture policies followed by the government of India from 1950-51 till 2000have passed three phases. The 1st phase was in the pre-green revolution or traditional agriculture period (1950-51 to mid-1960) where enormous agrarian reforms, development of major irrigation projects and institutional changes were made. The intermediary landlordism was abolished; tenant operations were given security of farming and ownership of land; land ceiling acts were imposed by different states to eliminate large sized holdings and cooperative credit institutions were strengthened to minimize exploitation of cultivators by private money lenders and traders (Radhakrishna 1993). Expansion of area was the main source of growth in the pre-green revolution period. Land consolidation was also affected to reduce the number of land fragments. This is a technologically stagnant phase in which a larger farm production becomes generally possible only through increased application of all three traditional inputs like land, labour and capital. Even if some elements of dynamic agriculture like application of fertilizer, improved seeds and land reform are introduced, the increase in productivity is smaller till the mid-1960s, Indian agriculture was typically embodied within the framework of traditional agriculture. The country faced severe food shortage and crisis in early 1960s which forced the policy makers to realise that continuous reliance on food imports and aid imposes heavy costs in terms of political pressure and economic instability (Rao 1996) and there was a desperate search for a quick breakthrough in agricultural production.

In the 2nd phase, the landscape of Indian agriculture underwent a transformative shift with the advent of the Green Revolution in the 1960s, ushering in a new era characterized by technologically driven farming practices that emphasized efficiency and productivity over extensive capital investment. This is the beginning of the process of transformation from traditional agriculture to modernisation. In this phase, agriculture still represents a large portion of the total economy. During this phase of the Green Revolution (1960-1980), the primary driver of crop output growth shifted towards increased productivity. This period witnessed a significant surge in yield growth, marking a notable acceleration in agricultural productivity. Consequently, wheat and rice production in a short span of 6 years between 1965-66 and 1971-72 witnessed an increase of 30 million tonnes, which was 168 percent higher than the achievement of 15 years following 1950-51(Ramesh Chand, India's National Agriculture Policy: A critique). So, there was self-sufficiency in food grains. But the scope for area expansion diminished considerably in the green revolution period in which the growth rate in area was less than half the growth rate in the first phase. Agrarian reforms during the green revolution were not given priority as compared to research, extension, input supply, credit, marketing, price support and spread of technology. Two important institutions like the Food Corporation of India and Agricultural Prices Commission were created in the beginning of the green revolution to ensure remunerative prices to producers, maintain reasonable prices for consumers, and to maintain buffer stock. These two institutions have mainly benefited rice and wheat crops. In this phase the dry land areas have not seen the benefit of technological breakthrough as compared to irrigated areas.

The 3rd Phase began in early 1980 and continued till formulation of 1stNational Agriculture Policy, 2000. This was the post green revolution phase. This phase marked as technologically dynamic agriculture with high capital intensity. This was precisely the period when the non-agricultural sectors also began their march towards modernisation. Non-agricultural sectors were facilitated in their move towards aggressive modernisation by the new policies of liberalisation, privatisation and globalisation. This phase of agricultural transformation is thus characterised by the substitution of labour by capital by way of large-scale farm machinery. Government initiated economic reforms in1991, which involved deregulation, reduced government participation in economic activities, and liberalization. Much of the reforms did not affect the agriculture sector. But agriculture was affected indirectly by devaluation of the exchange rate, liberalization of external trade. Then, came a new international trade accord and WTO, requiring opening up of the domestic market. There has been a considerable increase in subsidies and support to the agriculture sector during this period while public sector spending in agriculture for infrastructure development started showing decline in real terms but investments by farmers kept on moving on a rising trend (Mishra and Chand, 1995; Chand, 2001). The output growth, which was concentrated in very narrow pockets, became broad-based and got momentum. The rural economy started witnessing a process of diversification which resulted in fast growth in non-food grain output like milk, fishery, poultry, vegetables, fruits etc which accelerated growth in agricultural GDP during the 1980s (Chand, 2003). This growth seems largely market driven.

National Agriculture Policy, 2000 & subsequent initiatives-India's inaugural National Agriculture Policy in July 2000 aimed to unlock agricultural potential, bolster rural infrastructure, foster value addition, propel agribusiness, generate rural employment, elevate farmers' living standards, curb urban migration, and address challenges of economic liberalization and globalization. Key objectives included achieving over 4% annual output growth, ensuring inclusive growth, and fair distribution of benefits. Policies focused on augmenting food grain production, improving micro-level planning, water management, technology missions for various crops, rural warehouse expansion, cooperative policies, credit reforms, agricultural infrastructure, insurance schemes, legislation for plant varieties, seed policies, agricultural extension via mass media, Kissan Call Centres, agri-clinics, agribusiness centers, drought management, assistance to sugarcane farmers, and enhancing milk, egg, and fish production, alongside advancing agricultural research.Since the early 2000s, the Indian agricultural sector has significantly shifted towards sustainable practices, embracing organic farming and integrated pest management to address environmental concerns. Efforts have been concentrated on improving soil health, conserving water resources, preserving biodiversity, and adopting climate-resilient agricultural techniques. The Indian Council of Agricultural Research (ICAR) is leveraging Artificial Intelligence (AI) in agriculture to enhance productivity, efficiency, and sustainability. AI technologies like machine learning, remote sensing, and data analytics are being utilized to optimize crop management, predict weather patterns, manage pest outbreaks, and improve soil health. Overall, while there might not be a standalone drone policy specifically for agriculture, India is likely to integrate agricultural applications into its broader drone policy framework as part of efforts to modernize the agricultural sector and improve productivity.The integration of technology, such as precision agriculture and remote sensing, has been prioritized, complemented by initiatives like the Pradhan Mantri Krishi Sinchayee Yojana (PMKSY) aimed at enhancing irrigation capabilities at field level. To promote organic farming, the government has introduced schemes like the Paramparagat Krishi Vikas Yojana (PKVY) and the National Mission for Sustainable Agriculture (NMSA).Moreover, the Pradhan Mantri Fasal Bima Yojana (PMFBY) provides crop insurance coverage and financial aid to farmers in the event of crop failure due to natural calamities, pests, or diseases. The Pradhan Mantri Annadata Aay Sanrakshan Abhiyan (PM-AASHA) is a government initiative aimed at ensuring fair prices for farmers' produce. It coordinates with state governments through schemes like the Price Support Scheme (PSS), which provides a safety net for farmers by guaranteeing a minimum support price (MSP), the Price Deficiency Payment Scheme (PDPS), which compensates farmers for the difference between MSP and market prices, and the Pilot of Private Procurement and Stockist Scheme (PPPS), encouraging private sector involvement in procurement to enhance efficiency and market access for farmers.As part of the Atmanirbhar Bharat Abhiyan economic stimulus package, the Agriculture Infrastructure Fund was initiated to finance projects for post-harvest management and marketing of agricultural produce. Launched in 2016, the National Agricultural Market (eNAM) aims to establish a unified national market for agricultural commodities by integrating existing APMC mandis.The Pradhan Mantri Kisan Samman Nidhi (PM-KISAN) scheme, launched in February 2019, offers direct income support of Rs 6,000 annually to small and marginal farmers. India's digital policy in agriculture is multifaceted, aiming to leverage technology for enhanced farming efficiency and sustainability. It includes initiatives like expanding broadband access, safeguarding data privacy, and promoting interoperability (the ability of computer systems or software to exchange and make use of information standards). Essential components involve training farmers in digital skills, funding research, and ensuring cybersecurity. In a significant development, the Government of India promulgated three farm laws in 2020 aimed at facilitating a framework wherein farmers could directly sell their produce outside the purview of Agriculture Produce Market Committees, engage in contract farming, and exercise greater freedom in marketing their agricultural output. These laws also deregulated trade in essential commodities such as food grains, pulses, edible oils, and onions, barring exceptional circumstances. However, in response to widespread protests by farmers across the country, the Government of India repealed these contentious farm laws in 2021.

India's agricultural policies prioritize modernizing small-scale farms, engaging youth in agriculture, and promoting cooperative models. Aligned with the "Sabka Saath Sabka Vikas" ethos, these initiatives recognize agriculture's pivotal role, sustaining over half of India's workforce. Collaboration with the FAO shapes policies towards 2030, emphasizing sustainable development through national dialogues and directives. This vision aims for inclusive growth, enhanced productivity, and resilience in the face of evolving challenges, ensuring agriculture's continued contribution to India's prosperity.

Current status of Agriculture & allied sectors in India and Eastern States: Presently production of food grains increased over 6 times, horticultural crops by 13 times, fish by 18 times, milk by 10 times and eggs by 53 times since 1950-51, thus making a visible impact on the national food and nutritional security (Pathak and Ayyappan 2020). Currently India is the largest producer of cotton, pearl millet, pulses, jute, banana, papaya, mango, ginger, okra and milk; second largest producer of rice, wheat, sugarcane, groundnut, rapeseed, mustard, fruits, vegetables, cashew, potato, onion, cauliflower, brinjal and cabbage and third largest producer of citrus in the world on account of active efforts and contributions of agricultural scientists. It is also one of the leading producers of spices, fish, poultry, livestock and plantation crops. However, Indian agriculture continues to battle several intimidating challenges of increasing productivity, profitability and resilience at the backdrop of increasing population, depleting natural resource base, aggravating climate change and reducing farm income.

Field crops: In the years of early independence, the country faced a severe crisis of food to feed its 300 million people and was forced to depend on PL 480 of the USA. At that time, the country imported wheat from the United States of America on concessional prices to feed its hungry millions. During the last half of century (1965-2015) India's food production multiplied by 3.7 times while the population multiplied by 2.35 times. Thereby India has achieved food sufficiency and presently the world's largest rice exporter. Green Revolution started in mid1960s for increase of food grain production specially paddy & wheat through high yield varieties, other technologies and irrigation development. Nobel Laureate Dr Norman E. Borlaug, Dr M.S. Swaminathan and other Indian scientists guided Indian farmers to adopt dwarf varieties of wheat that started Green Revolution in India during 1960s.

Yellow Revolution launched in 1986-87 for increase the production of edible oil, especially mustard and sesame seeds to achieve self-reliance, Gene Revolution launched for application of biotechnology in food production in 2002 to help the increase of the crops yield through introducing high-yielding varieties resistant to biotic and abiotic stresses, reduce pest associated losses & increase the nutritional values of foods. Pulse Revolution launched in 2009-2010 for increase of pulses. The agricultural & allied sector production from 135 million tonnes in 1950-51 was increased to over 1300 million tonnes in 2021-22 (ICAR, 2022, Indian Agriculture after Independence publication).

Food grains: India moved from food-aid dependency to self-sufficiency in food production during last 75 years of independence. Between 1951 and 2021, food grain production increased six-fold. Such achievements became possible with the innovations in technologies including high-yielding varieties, irrigation, fertilizer, and farm policies. In India the food grain production increased from 50.82million tonnes out of an area of 30.81 million ha in 1950-51 to 315.72 million tonnes out of an area of million ha during 2021-22. The food grains production (in million tonnes) in major food growing states of India during 2021-22 are Uttar Pradesh (56.11), Madhya Pradesh (39.05), (Punjab 28.21), Rajasthan (21.04), Maharashtra (17.13), Haryana (16.32) and Bihar (16.19). The food grain production (million ha) in eastern Indian states is West Bengal(20.50), Bihar (16.19) , Odisha (9.95), Chhattisgarh (8.89) , Assam (5.53) and Jharkhand (4.98) during 2021-22.

Rice: India is the second-largest rice-producing country contributing 22% of the world's production. In India the rice production increased from 20.58 million tonnes out of an area of 97.2 million ha in 1950-51 to 130.29 million tonnes out of an area of 46.40 million ha during 2021-22. India as a partner of the global initiative contributed to solve the rice genome (genetic information). Complete genetic code is helping rice breeders to quickly develop varieties with specific traits such as stress resistance and high yield. The rice production (in million tonnes) in major rice growing states in India are West Bengal (16.76), Uttar Pradesh (15.27), Punjab (12.88), Telangana (12.30), Odisha (9.13), Tamil Nādu (8.06) and Andhra Pradesh (7.78). Among eastern Indian states, West Bengal contributed 12.8%, Odisha 7%, Chhattisgarh 6%, Bihar 5.8% and Assam 4% of India's rice production during 2021-22. India, the largest rice exporter in the world, shares about 90% of Basmati rice trade in overseas market. The country exported 17.73 million tonnes of rice earning Rs 65,326 crore in 2020-21(Dr SK Malhotra Project Director, Directorate of Knowledge Management in Agriculture, ICAR (2022), ICAR Transforming Indian Agriculture.

Wheat: The wheat production has increased of over 16 times from 6.46 million tons in 1950-51 to 109.52 million tons in 2020-21. India is the second largest wheat producer in the world. Wheat served a pertinent role in progress of the nation towards economic growth as well as food and nutrition security. Wheat varieties developed over the years by extensive breeding for higher yield, superior quality traits and resistance to climate adversities led to self-sufficiency. The year 2020, saw record production as well as procurement of wheat. It provided the required buffer stock that have the capacity to feed millions of the Indians free of cost during the tough times. The wheat production (in million tonnes) in major wheat growing states in India during 2021-22 are Uttar Pradesh (33.94), Madhya Pradesh (22.41), Punjab (14.82), Haryana (10.44), Rajasthan (9.48) & Bihar (6.22).

Maize: Maize is important to India as 15 million Indian farmers are engaged in maize cultivation. Maize qualifies as a potential crop for doubling farmer's income. In India, maize is the third most important food crops after rice and wheat with an annual output of 31.5 million tonnes. Maize is a multi-faceted crop used as food, feed and industrial crop globally. Bulk of the maize production in India (approximately 47%), is used as poultry feed. Of the rest of the produce, 13% is used as livestock feed and food purpose each, 12% for industrial purposes, 14% in starch industry, 7% as processed food, and 6% for export and other purposes. Maize can also be consumed as specialized maize like quality protein maize (QPM), baby corn, sweet corn etc. It is a source of nutrition as well as phytochemical compounds. Phytochemicals play an important role in preventing chronic diseases. It contains various major phytochemicals such as carotenoids, phenolic compounds, and phytosterols. It is believed to have potential anti-HIV activity due to the presence of Galanthus nivalis agglutinin (GNA) lectin or GNA-maize (Tajamul Rouf Shah *et al.*, 2016). Maize is grown in two seasons, rainy (kharif) and winter (rabi)in India. Kharif maize represents around 83% of maize area in India, while rabi maize correspond to 17% maize area. Over 70% of kharif maize area is grown under the rainfed condition. The stress prone ecology contributes towards lower productivity of kharif maize (2706 kg/ha) as compared to rabi maize (4436 kg/ha), which is predominantly grown under assured ecosystem. Among cereals maize has highest growth rate in terms of area and productivity. Since 2010 maize productivity in India is increasing @ over 50 kg/ha/year, which is the highest among food crops. In India maize production was 31.51 million tonnes from an area of 9.86 million ha with productivity of 3195 kg/ha during 2020-21. The maize production (in million tonnes) in major maize growing states in India during 2020-21 are Karnataka (5.18), Madhya Pradesh (3.58), Maharashtra (3.44), Tamil Nādu(2.72), West Bengal (2.45), Rajasthan (2.27), Bihar (2.22), Andhra Pradesh (1.95) and Telangana (1.75).

Millets: Most of the millet crops are native of India and are popularly known as Nutri-cereals as they provide most of the nutrients required for normal functioning of human body. Ministry of Agriculture and Farmers Welfare, GOI has recognized the importance of millets and declared millets comprising of sorghum, pearl millet, finger millet, minor millets (foxtail millet, proso millet, kodo millet, barnyard millet, little millet (Kutki) and two pseudo millets (buck-wheat (Kuttu) and Amaranthus (chaulai) as "Nutri-Cereals" for production, consumption and trade point of view. 154 varieties of different crops of millets have been released in between 2014-2021 in India. (Source: Presentation for International year of millets, 2023 in National Conference on Kharif Campaign, 2022 on 19th April, 2022 by Ministry of Agriculture & Farmers Welfare). In India, millets are grown in an area of 13.8 million hectare with production of 17.3 million tonnes. The millets production contributes about 10% to the country's food grain requirement. But in the past six decades, the area under millets has gone down by 62.57%, dropping from 36.34 million ha during 1955-56 to around 13.83 million during 2019-20 (Indian Agriculture after Independence, ICAR, 2022). United Nations declared 2023 as international year of Millets as per proposal of India.

Odisha Millet Mission: (OMM) is a flagship programme of the government of Odisha launched in 2017 to improve nutrition at the household level through the revival of millets in farms and on plates of tribal communities in Odisha. World Food Programme, India, signed an MoU with Odisha Millet Mission (OMM) to assess the key achievements of OMM, the operational model followed, plans, lessons learned and documents the best practices of OMM experience to develop a range of knowledge products. The assessment included the end-to-end value chain interventions under OMM and is identifying scalable and innovative approaches. OMM provides farmers incentives of up to Rs 9,500 per ha over a three-year period to shift to millet cultivation. This includes free seeds and organic fertilizers. To complete the supply chain, the government of Odisha has created an assured market for the produce. The state has been offering minimum support price of Rs 3,377 per 100 kg of ragi. The government of Odisha has also made it part of the public distribution system (PDS) and provides 1 kg of ragi per ration card holder, which can increase to 2 kg in areas with high ragi procurement. Over the past five years, Odisha has recorded millet production, from 3,333 hectares (ha) in 2017-18 to 53,230 ha in 2021-22. Even the average yield has improved by 28 per cent because of sustainable agricultural practices and quality seeds. One of the key interventions under OMM is promotion of value addition enterprises through women SHGs under Mission Shakti. Millet Shakti outlets have been established in different districts of the state to sell variety of millet-based foods products. By using millets, SHGs prepare dosa, samosa, boondi, laddu, mudik, pakoda, halwa, kheer, cookies, upma, pulao etc.

QUINOA (prounced as Keen-wah)- A super food crop under water scarcity: Considering the scope and prospects of Quinoa cultivation, the United Nations declared 2013 as the International Year of Quinoa.Government of India has recognised the importance of quinoa as a climate resilient crop and included under Millet Mission as a pseudo millet crop for its wide adoption. Quinoa (Chenopodium quinoa), Amaranthaceae family originally from the Andean mountains of South America, is adaptable to different types of soil and climatic conditions. Like amaranth seeds quinoa seeds are also categorized under pseudocereals as these grains are not from grasses like wheat and rice. The crop can be cultivated in extreme climate and weather condition. Ideal temperature for quinoa cultivation is around 18-20°C, however it can withstand temperature extremes ranging from 36°C to -8°C. Research reported that temperatures which exceeded 36°C tended to cause plant dormancy or pollen sterility. This crop is very well responding to climate when sown duringrabi season (mid October-mid December) in India, contributing to higher seed yield. Quinoa, surpassing major cereals, minor millets, and pulses in nutritional value, has the potential to serve as a viable substitute for cereals in the near future. Quinoa seed is gluten free &rich in proteins, lipids, fibres, Ca, P, Mg, Fe, Zn, Na and Cu. All the nine amino acids that are strictly essential for humans are present in quinoa.Quino contains 4.2 % fibre which is more than rice, wheat, corn, pearl millet, oat & barley. The comparisons of the nutritional quality of quinoa with various grains is given at Table 1.

Table 1. Comparisons of the nutritional quality of quinoa with various grains.

Crops	**Calcium (mg)**	**Carbohydrate (gm)**	**Protein (gm)**	**Fat (gm)**	**Iron (mg)**	**Energy (kcl)**
Quinoa	47.00	61.86	15.36	6.00	4.70	399
Barley	8.60	77.72	9.91	1.16	2.00	336
Pearl millet (Bajra)	38.00	64.00	10.60	4.80	8.00	363
Foxtail millet	31.00	60.90	12.30	3.30	2.80	351
Sorghum (Jowar)	15.00	74.63	11.30	3.30	4.00	339
Oats	11.00	67.00	11.00	6.30	4.00	384
Wheat	29.00	71.28	12.61	1.54	3.19	327
Rice (Brown)	11.00	77.24	7.94	2.92	1.60	370
Corn	7.00	74.26	9.42	4.74	2.70	365

Source: FAOSTST (Quoted by K. Srinivas *et al.* in their paper titled" Analysing the Value Chain of Quinoa: A Case Study of Quinoa")

Quinoa is claimed to reduce the risk of various diseases in human beings like cardiovascular diseases, type-2 diabetes, high blood pressure and obesity. Although the quinoa seeds are very nutritious, the same must be processed to remove the coating containing the bitter tasting saponin after harvest. The seeds are in general cooked the same way as rice and can be used in a wide range of dishes. The leaves are eaten as a leaf vegetable, much like amaranth. The byproduct – saponin can be used for liquid soap, jewellery polish, detergent, eczema /dermatitis cure, pesticide/insecticide, pet shampoo, human shampoo, surf spreader/sticker, antimicrobe, parasite remover (tick, flea) etc.Quinoa is a relatively new crop in India, its craze among the urban people is increasing day by daydue to nutritional quality of the seeds.In recent years the crop is becoming popular among the farmers in India due to its drought resistance.The crop is grown in isolated areas of Rajasthan, Himachal Pradesh, Karnataka, Tamil Nadu, dry tracts of Andhra Pradesh. Recently, a significant area of production was also noted in Rajasthan, Uttarakhand and Uttar Pradesh. The exact area and production are yet to come out. The cultivation of quinoa is being subsidised by Govt. of Rajasthan to promote more production in the state.The food and nutritional value along with well accepted agronomic characters of quinoa crop has opened opportunities to cultivate in various states of India. The crop canbe fitted to rice fallows of eastern India with minimum water and pest management practices particularly true for Odisha. During wet season, rice is leading cereal crop in Odisha. About 2.1 million hectares rice fallows are underutilised due to lackof assured irrigation facilities. Short duration and low water requirement crop like quinoa can be taken with minimum effort in rice fallows of several districts of Odisha (B S Satapathyetal, ICAR-Indian Institute of Water Management, Bhubaneswar, Quinoa: A climate resilient crop, Indian Farming 73 (05): 22-24; May 2023). Average yield of quinoa1.0–1.5 t/ha is generally reported under field conditions. Indian quinoa market size reached 62,263 tons in 2022 as per market research covering Rajasthan, Andhra Pradesh, Uttar Pradesh, Maharashtra, Himachal Pradesh, Tamil Nadu, Madhya Pradesh, Uttarakhand, Haryana, other states by IMARC (International Market Analysis Research and Consulting) group (Ref: IMARC, Market Research Report ID: SR112023A3292). Looking forward, IMARC Group expects the market to reach 232,991 tons by 2028, exhibiting a growth rate (Compound Annual Growth Rate) of 24.7% during 2023-2028.

Pulses: Availability of new varieties of pulses and large-scale field demonstrations made farmers aware about the seeds of new varieties of pulses. Seed hubs on pulses established across the country provided quality seeds to farmers. Continued efforts for about three decades in pulses brought 'Pulse Revolution' by providing near self-sufficiency in pulses in the late 2010s

(Yadav *et al.* 2019). Mission mode approach revolutionized pulse production, broke yield barriers and helped India to reduce import-dependence. The production of pulses, which remained almost constant at 14-15 million tonnes till 2014-15, jumped to 27.69 million tonnes in 2021-22. The pulses production (in million tonnes) in major pulses growing states in India during 201-22 are Madhya Pradesh (6.03), Maharashtra (5.19), Rajasthan (4.01) ,Gujarat (2.68), Utter Pradesh (2.56), Karnataka (1.96) and Andhra Pradesh (1.08).

Oilseeds: Varietal improvement and complementary technologies for oilseeds have been developed. More than 800 new varieties of mustard, groundnut, soybean etc. were released in the country. Quality seeds were produced in seed hubs and large-scale demonstrations in farmers' field are contributing to the edible oil economy of the country The edible oil production from 10.8 million tonnes (1985-86) was increased to 24.7 million tonnes (1998-99) due to adoption of various technologies under the Yellow Revolution in the country. The oilseeds production in India increased from 8.56 million tons out of an area of 15.26 million ha during 1964-65 to 37.69 million tons out of area of 29.17 million ha during 2021-22. The oilseeds production (in million tonnes) in major oilseeds growing states in India during 2021-22 are Rajasthan (8.38), Madhya Pradesh (7.92), Gujarat(6.89) and Maharashtra (5.89).

Sugarcane: Sugarcane, a major cash crop, is providing employment to over a million people directly or indirectly. India is number one in sugarcane productivity among major sugarcane growing countries. Sugarcane hybrid (Co 86032), a dominant one in Maharashtra covering more than 50% area, is tolerant to drought, smut and red rot. Improved sugarcane varieties have made a major contribution in increasing yield. Besides improved agro-practices, and low-cost plant protection measures helped record sugar production. It made India not only self-reliant in sugar production but also an exporting country. India witnessed Sugar Revolution with sugarcane production increasing from 57.05 million tonnes out of an area of 1.71 million ha in 1950-51 to 431.81 million tonnes out of 5.15 million ha area in 2021-22. Sugarcane production (in million tonnes) in major sugarcane growing states in India during 2021-22 are Uttar Pradesh (177.42), Maharashtra (110.53), Karnataka (61.15), Gujarat (17.45), Tamil Nadu (15.45) and Bihar (12.06).

Cotton & Jute and Mesta: Till 1970s, country used to import massive quantities of cotton in the range of 8.00 to 9.00 lakh bales per annum. India developed the first cotton hybrid (H-4) in the world during 1970. Since then, country has become self-sufficient in cotton production barring few years in the late 90s and early 2020s when large quantities of cotton had to be imported due to lower crop production and increasing cotton requirements of the domestic textile industry. India is unique in terms of growing all the four cultivated species of cotton. Pest-resistant Genetically Modified (GM) Bt cotton hybrids have captured the Indian market (covering over 90% of the area under cotton) since their introduction in 2002. India accounts about 36-38% of the global cotton acreage and 25% of production. India produced 31.20 million tons of cotton(lint) from an area of 11.91 million ha area with a productivity of 445 kg per ha during 2021-22. Cotton (lint) production (in million tons) in major cotton growing states in India are Gujrat (7.48), Maharashtra (7.11), Telangana (6.06), Rajasthan (2.48), Karnataka (1.95), Andhra Pradesh (1.70), Madhya Pradesh (1.41) and Haryana (1.31).

India produced 10.31 million tons of Raw Jute & Mesta from an area of 0.69 million ha area with a productivity of 2685 kg per ha during 2021-22. The raw Jute and Mesta production (million tons) in major Jute growing states are and West Bengal (8.35) , Assam (0.90) and Bihar (0.81).

Production (in million tonnes) of Rice, Wheat, Pulses, Food grains, Oilseeds, Cotton (lint), Sugarcane, Raw Jute & Mesta in major eastern Indian states & comparisons with India (Data are based on Fourth Advance Estimates, Ministry of Agriculture & Farmers Welfare, Government of India) are given in Table 2.

Table 2. Production (in million tonnes) of Rice, Wheat, Pulses, Food grains, Oilseeds, Cotton, Sugarcane, Raw Jute & Mesta in major eastern Indian states & comparisons with India during 2022-23 (Data for 2021-22 are based on Fourth Advance Estimates, Ministry of Agriculture & Farmers Welfare, Government of India)

State	Rice	Wheat	Pulses	Food grains	Oil Seeds	Cotton (lint)	Sugar cane	Raw Jute & Mesta
Assam	5.268	0.017	0.118	5.535	0.198	0.0005	1.117	0.905
Bihar	7.064	6.220	0.363	16.194	0.114	-	12.063	0.819
Chhattisgarh	7.897	0.186	0.387	8.897	0.104	0.0075	1.990	0.0009
Jharkhand	2.927	0.507	0.898	4.984	0.398	-	-	-
Odisha	9.136	0.0003	0.462	9.950	0.146	0.626	0.397	0.067
West Bengal	16.762	0.661	0.427	20.502	1.192	0.0008	1.518	8.350
India	130.290	106.849	27.692	315.722	37.696	31.20	431.811	10.310

Source: Hand Book of Statistics on the Indian Economy, 2021-22, Reserve Bank Of India.

Horticulture: India gave priority for increase of production in staple crops like paddy and wheat for self-sufficiency in food after independence. The green revolution from 1960 to1980 made food security in the country. Thereafter till 1990 the gains of green revolution were plateauing and the area expansion under field crops were gradually diminishing. At this stage the planners, and scientists searched for an alternate source of growth in the agriculture sector. The high value horticultural crops were thought to be an important source of growth of farm income. Government of India gave priority by allocating Rs 1000 crores for this sector in VIII five-year plan from 1992-97. This policy of the government helped this sector to achieve a compound annual growth rate of 4.82% which is above the target 4% annual growth rate fixed in the National Agriculture Policy, 2000. The period between 1991-2003 was regarded as the golden revolution period because, during this period, the investment planned in the horticulture segment became highly productive. India became the world leader in the production of a variety of fruits like coconut, mangoes, cashew nuts and more.

Horticultural crops: In India the production of horticultural crops increased 13-fold from 25 million tonnes during 1950-51 to more than 331 million tonnes during 2020-21, which has surpassed the food grain production (Ministry of Agriculture & Farmers' Welfare, GOI, 2021). At present horticultural crops are cultivated in 18% of the total cultivated area, but contribute 33% of GVA in agriculture GDP. Our country occupies 2nd position in the world under total production of horticultural crops next to China. India contributes 11.4% of world's total horticulture production and 11.8% of total fruits and vegetable production. India could get Rs. 9941 crore from export of fruits and vegetables during 2020-21 in spite of the corona pandemic. Now horticulture has got a place in the Integrated Farming System models for doubling farmers' income. The production (in million tonnes) of horticultural crops in major growing states of India during 2020-21 are Uttar Pradesh (40.82), Madhya Pradesh (33.32), West Bengal (32.74), Maharashtra (27.78), Andhra Pradesh (26.52), Gujarat (23.54), Bihar (22.33) and Karnataka (20.38)- *Source:* Agricultural Statistics at a glance, 2021, GOI.

Fruits: India is the second largest producer of fruits in the world with amazing diversity across the country. India is the world leader in the production of mango, banana and papaya. Mango is an important fruit crop in India. Amrapali, a popular hybrid of famous 'Dussheri' and 'Neelam' varieties. It is dwarf, suitable for high-density plantation and a regular bearer with clusters of small-size fruits. The citrus with shoot-tip grafting producing virus free planting materials give higher yield. The Bhagya variety of Pomegranate gives a very high income to the farmers. India is the largest producer of pomegranate in the world. Scientists standardized the cultivation of grapes on dog ridge rootstock (mitigate salinity and drought problems), which resulted in improved quality of grapes, reduced production cost, increased yield and higher profits for farmers. Nearly 90% of grape plantation is based on the rootstock. Export of bananas to

Italy and the Middle East was initiated. During 2020-21, India exported fruits worth Rs 4,971.22 crore. Grapes, pomegranates, mangoes, bananas, oranges account for a larger portion of fruits exported from the country. (Source: Dr SK Malhotra Project Director, Directorate of Knowledge Management in Agriculture, ICAR, 2022, ICAR Transforming Indian Agriculture). As per agricultural statistics at a glance,2021, GOI, the production of fruits (million tons) in major states of India are Andhra Pradesh (18.10) Maharashtra (12.84), Utter Pradesh (11.23), Madhya Pradesh (8.24) and Gujarat (8.19). Similarly, the production of fruits (million tons) in eastern Indian states are Bihar (4.63), West Bengal (3.65), Odisha (2.78), Assam (2.51), Chhattisgarh (2.46) and Jharkhand (1.29).

Mango (King of Fruits): Mango is an economically significant fruit & rich in carbohydrates, minerals, dietary fibre (pectin), vitamin C, and vitamin A (β-carotene) and several other phytochemicals. India exported 27,872 tonnes of fresh mangoes to different countries of world for the worth of Rs 327.45 crores during 2021-22 (APEDA). A few hybrids like Amrapali, Mallika, Ambika and Arunika, are being cultivated in various agroecology around the country. But excellent hybrid like Ratna is limited to Alphonso growing belt of Maharashtra. Amrapali is well known for its regular bearing and excellent performance in areas with good humidity, particularly in coastal parts of West Bengal and Odisha. Alphonso, Dusseheri, Kesar, and many other mango varieties are famous around the world for their unique taste, distinct flavour, and aroma. The GI tag serves as proof of a product's quality and unique identity. The different certified varieties of mango from different geographical areas are Laxmanbhog, Himsagar, Fazli, MalihabadiDusseheri, Banganpalle, Alphaonso, Ratahi, Marathwad Kesar, Gir Kesar, Salem mango etc. Presently Chausa and Langada are under process of registration for getting GI. tag. India produced 20.33 million tonnes of mango from an area of 2.33-million-hectare area as per 1st advance estimates of 2021-22 (agricoop.nic.in). The mango production in India from major mango growing areas as per 1st advance estimate of 2021-22 are Utter Pradesh (4.80), Andhra Pradesh (4.67), Karnataka (1.74), Bihar (1.54) and Telangana (1.15).

Cashew nut: This crop is an important plantation crop in India and the major growing states are Maharashtra, Andhra Pradesh, Odisha, Karnataka, Tamil Nadu and Kerala. India produced 0.77 million tonnes of cashew nuts from an area of 1.16 million hectares area as per 1st advance estimates of 2021-22 (agricoop.nic.in).The production of cashew nut (million tons) in major cashew nut growing states are Maharashtra (0.199), Andhra Pradesh (0.127), Odisha (0.121), Karnataka & Tamil Nadu each (0.077), and Kerala (0.076) as per

1st advance estimates of 2021-22 (agricoop.nic.in). The highest productivity of cashew nuts in India was 946 kg/ha from Maharashtra followed by 705 kg / hain Kerala. Cashew nut hybrid H 130 starts setting seeds in the first year itself, while others fruit in the second or third year. For the first time in India, such a dwarf cashew genotype, to reduce pruning in high density planting systems, was approved for release. Overall, the country has 3940 cashew processing units, with a total installed capacity of 1643 thousand tonnes and an average installed capacity of 0.4 thousand tonnes. Maharashtra has the majority of them with 55.8%. Kerala, on the other hand, has the greatest installed capacity of 36.5%. In eastern Indian states, Odisha has 350 processing units with an installed capacity 12 thousand tonnes, West Bengal 30 processing units with an installed capacity 8 thousand tonnes and Chhattisgarh 3 processing units with an installed capacity 5 thousand tonnes (International Journal of Case Studies in Business, IT, and Education (IJCSBE), ISSN: 2581-6942, Vol. 5, No. 2, December 2021).

Coconut: The coconut crop plays a very significant role in the agrarian economy of India. There are about five million coconut farmers in India and most of them are small and marginal. Coconut industry provides livelihood to about ten million people in India. Coconut trees are called "Kalpavriksha"which essentially means that all parts of a coconut tree are useful in one way or the other. Tender coconut water contains sugar, simple proteins, amino acids like alanine, cysteine & serin, fatty acids like oleic acid& linoleic acid, minerals like K, Na, Ca, P, Fe, Cu, Mg etc and gives relief to number of diseases like urinary problems, stroke, blood pressure , measles ,dehydration etc. Other value-added products of coconut are coconut sugar, coconut chips, dark chocolate, drinking chocolate, frozen delicacy etc. India produced 13.65 million tonnes of Coconut from an area of 2.22million hectares area as per 1st advance estimates of 2021-22 (agricoop.nic.in). The production of coconut (million tons) in major coconut growing states(including eastern states)are Karnataka (4.21), Tamil Nadu (3.75), Kerala (3.30), Andhra Pradesh (1.12), est Bengal (0.278) and Odisha (0.273). There exists a huge scope for coconut based agri-business in India in order to increase the present 8% level of value addition to 25%, thereby value added products becoming a deciding factor in the price movement of coconut to ensure fair, reasonable and steady price to coconut farmers. Foreseeing the imperativeness of high value coconut sector, ICAR-CPCRI has developed complete package of practices for the production of virgin coconut oil (hot and fermentation process), coconut chips, coconut honey, jaggery and sugar. The Institute has also developed a technology for collecting coconut inflorescence sap by using a device. The sap thus collected is called Kalparasa which can be preserved up to 45 days under cold condition (in refrigerator) without adding any preservatives and additives with the bottling technology. It has been demonstrated that a farmer tapping 15 coconut palms for Kalparasa could earn on an average Rs. 45,000 a month, while a tapper can earn about Rs. 20,000per month. For sustaining the value-added coconut sector, Women Self Help Groups can be formed and equipped with technical know-how and smooth functioning of the coconut value chain can be ensured through continuous supply of value-added products to the downstream part of the chain.

Vegetables: India is the second largest producer of vegetables in the world. India produced 199.88 million tonnes of vegetables from an area of 11.06-million-hectare area during 2021-22, 1st advance estimates (agricoop.nic.in). The vegetables production (in million tonnes) in major vegetables growing states in India are Uttar Pradesh (29.58), West Bengal (28.22), Madhya Pradesh (20.59), Bihar (17.76), Maharashtra (16.78), Gujarat (15.68), and Odisha (9.52). The major vegetables production (in million tonnes) in India during 2021-22 are Potato (53.60), Onion (31.12), Tomato (20.30), Brinjal (12.76), Cabbage (9.60), Cauliflower (9.28) and Okra(6.41) as per 1st estimate during 2021-22(agricoop.nic.in). The per capita vegetable availability in India is approximately 400 g/day. High yield potential has been combined with multiple-disease resistance in hybrids of tomato, namely Arka Rakshak, Arka Samrat and Arka Abhed. India is the second largest producer of onion and garlic in the world. Indian onion is a favoured commodity in many African countries. In the last 2 decades, production of onion in India has increased more than four times and of garlic two times due to modern technologies and superior varieties developed with higher yield and quality. Development of disease and pest resistant varieties with matching production technologies led to a revolution in vegetable production. New vegetable varieties are being released from time to time. Mass production and distribution of quality seeds and planting materials of vegetables helped to raise productivity and income of farmers.

Potato: Potato is designated as "Food for Future" by FAO because of its high productivity and nutritive value. In India potato production & productivity have increased by 5.35 times &1.91 times respectively during 2020-21 as compared to 1978-79. (Agricultural Statistics at a glance, 2021, GOI). It has already emerged as the third most important food crop in the world after rice and wheat. India produced 53.60 million tonnes of potato from an area of 2.20 million hectares area as per 1st advance estimates of 2021-22 (agricoop.nic.in). However, it is projected that India needs to produce 55.00and 122 million tons of potatoes by 2025 and 2050, respectively to meet the demand of the growing population and economy (S. K. Chakrabarti, ICAR Sponsored winter school, 2019). Country's 27.66 % area and 29.16 % production of potato alone came from Uttar Pradesh during 2020-21. The second largest producing state of India, West Bengal produced 24.36 % of the country's production. Bihar produced 16.6 % of the country's potato production during 2020-21.

Onion: Onions have sharp, strong taste and smell due to its organic sulphur content. Organic sulphur compounds help to reduce the level of cholesterol in the body. Both the quercetin and organic sulfur compounds found in onions are known to promote insulin production, making them a helpful vegetable choice for those with diabetes. In India onion is grown under three crop seasons i.e., kharif, late kharif and rabi. Main crop is in rabi which covers about 50% of production whereas 20% is from kharif and 30% from late kharif. Maharashtra, Madhya Pradesh, Karnataka, Gujarat, Bihar, Rajasthan, Andhra Pradesh and West Bengal are the main onion growing states. India produced 31.12 million tons of onion from an area of 1.91 million hectares area as per 1st advance estimates of 2021-22 (agricoop.nic.in). Country's 42.95% area and 39% production of onion alone came from Maharashtra during 2020-21. The second largest onion producing state of India is Madhya Pradesh which produced 16.9% of the country's production during the same year. Among eastern Indian states, Bihar and West Bengal produced 4.9% and 2.79 % respectively of the country's onion production during 2020-21. About 90% of the export of onion is from Maharashtra. The productivity in late kharif and rabi is around 25 tons per hectare, whereas in kharif season it is 8-10 tons per hectare.

Brinjal: In our country Brinjal is the fourth most important vegetable after potato, onion and tomato. India produced 12.76 million tonnes of brinjal from an area of 0.74 million ha as per 1st advance estimates during 2021-22(agricoop. nic.in). The brinjal production (in million tonnes) in major brinjal growing states in India during 2021-22, 1st advance estimate are West Bengal (3.02) being the highest producing state from an area of 0.162 million ha, Odisha (2.12) being the second highest producing state from an area of 0.126 million ha, Gujarat ((1.53) from an area of 0.077 million ha, Bihar (1.20) from an area of 0.057 million ha and Madhya Pradesh (1.11) from an area of 0.052 million ha. The eastern Indian states namely West Bengal, Odisha and Bihar produced 49.68% of the country's production of brinjal during 21-22, 1st advance estimate. Brinjal in Odisha contributes 16 % of total area / total production of India. Odisha produces about Rs.75,800 crores worth of agricultural and allied output, out of which 6.8 per cent generated from brinjal (SAMRUDHI, Agriculture Policy,2020, Government of Odisha).

Tuber crops: The different tuber crops are sweet potato, taro, cassava, elephant foot yam and yams. The tuber crops are rich in dietary fibre, minerals, vitamins and bioactive phytochemicals. Tubers can sustain food, nutrition, health and livelihood security and help to mitigate the climate change effects. Tubers play a critical role in livelihood and food security of tribal farmers. Tuber starch finds its way in many industries. Suitable varieties developed for higher yields; and technologies developed for a wide range of value-added products–food to industries and from polymers to gels. One eastern Indian state, Odisha produced 0.381 million tons of sweet potato from an area of 40,410 ha during 2017-18. Odisha ranks No1 in area (30. 9 % of the country's area) and production (25.4% of the country's production) in India. Vitamin A deficiency increases the risk of night blindness. One possible solution for addressing vitamin A deficiency is through a food-based approach using orange-fleshed sweet potato as an inexpensive source of beta-carotene (the precursor to vitamin A). Elephant foot yam is rich in medicinal values. It is used for various ailments such as piles and constipation. Elephant foot yam is also having great flexibility to adopt in mixed farming systems. They can also be fitted into varied agro-ecological conditions. India produced 4.74 million tons of tapioca from an area of 0.13 million ha as per 1st advance estimates during 201-22 (agricoop.nic.in).83.73% of total production of tapioca in India was produced from Tamil Nadu. Similarly the production of sweet potato was 1.19 million tons from an area of 0.10 million ha in India. The production of sweet potato (million tons) from major sweet potato growing states are Odisha (0.33), Utter Pradesh (0.25) and west Bengal (0.18). In India elephant foot yam production was 0.81 million tons from an area of 0.034 million ha during 2021-22, 1st advance estimate. The highest production of elephant foot yam was 0.34 million tons in West Bengal followed by Andhra Pradesh 0.22 million tons.

Mushroom: The production of mushroom, a crop of waste to wealth, has almost doubled (from 1.00 lakh tonne to 2.58 lakh tonnes) by 2019 in India in less than a decade. (ICAR-Directorate of mushroom Research, Solan, Annual Report, 2019). The production of mushroom was 2.588 lakh tons during 2021-22, 1st advance estimates(agricoop.nic.in). Bihar produced 28,000 tons of mushroom, highest among the states of India followed by Maharashtra 25,600 tons & Odisha 25,00 tons. The high yielding varieties and farmer friendly technologies have played a significant role for such growth. Development of temperature-tolerant strains and in-house cultivation technologies pushed its cultivation to new heights. Popular in urban markets owing to nutritional and health benefits and now with new varieties, mushroom has become popular among rural masses.

Floriculture: Growth of floriculture is at an all-time high with increasing demand in domestic and overseas markets. Technologies to improve vase-life along with appropriate packaging technologies improved marketing and business potential. The area and production under flower crops almost tripled in the last two decades. Indian desi varieties of marigold have longer blooming period and good shelf-life which have been backed with improved cultivation technologies, turning fields into golden cash. Varieties of marigold suitable for different temperature regimes developed to ensure year-round cultivation. India produced 2.88 million tonnes of flowers including 0.79 million tonnes of cut flowers and 2.09 million tonnes of loose flowers from an area of 0.26 million tons during 202122, 1st advance estimates. In India, marigold production was highest out of the total production of flowers. India produced 0.74 million tonnes of marigold including 0.02 million tonnes of cut flowers and 0.72 million tonnes of loose flowers from an area of 0.07 million tons during 202122. Protected cultivation of commercial flowers in high-tech greenhouses standardized and extended to farmers. Technology for construction of low-cost polyhouse developed. North-eastern India is a treasure trove of orchids India produced 15,160 tons of orchids from an area of 2095 ha during 2021-22., 1st advance estimates. 91.88 % of orchid production in India comes from Assam state.

Honey: India exported honey worth Rs716 crore in the year 2020-21due to development of various technologies for Melipona culture, i.e. scientific domiciles for stingless bees, integrated management of swarming and absconding in Apis cerana, low cost technology for artificial domiciliation of solitary bees, etc. Honey production in the country during 2020-21 increased to 1.25 lakh metric tonnes from 76,150 metric tonnes in 2013-14.

The production (in million tonnes) of horticultural crops, fruits, vegetables, plantation crops and spices in major eastern Indian states & comparisons with India as per 1st Advance estimates of 2021-22, GOI (agricoop.nic.in). are given in Table 3.

Table 3. Production (in million tonnes) of horticultural crops, fruits, vegetables, plantation crops and spices in major eastern Indian states & comparisons with India as per 1st Advance estimates of 2021-22, GOI (agricoop.nic.in).

State	Horticultural crops	Fruits	Vegetables	Plantation crops	Spices
Assam	7.32	2.51	4.19	0.180	0.330
Bihar	22.33	4.63	17.57	0.054	0.014
Chhattisgarh	9.48	2.46	6.85	0.029	0.018
Jharkhand	5.03	1.29	3.73	0.006	-
Odisha	13.04	2.78	9.51	0.390	0.294
West Bengal	32.74	3.65	28.43	0.300	0.253
Odisha	331.04	103.02	197.23	16.60	10.69

Source: 1st Advance estimates of 2021-22, GOI (agricoop.nic.in).

Processing of agricultural produce: This is of primary importance as monetization of produce can be made into appropriate value. Processing can happen at three different levels, namely (i) primary processing (simple farm gate practices like cleaning, sorting, grading, packaging, etc); (ii) post-harvest secondary processing (basic processing, packaging & branding); (iii) high-end processing which involves complex processing technologies, machinery and finances, with the output of a rich range of products .Currently, commercial processing of fruits and vegetables is extremely low in India, at around 2.2% of the total production as compared to countries like Japan 81%, Philippines 78%, Malaysia 77%, United States (U.S.) 65% and China at 23%. It is significant to note that the current installed capacity can process only 3 to 4 per cent of the total production of fruits and vegetables in India. There were around 4100 to 4200 processing units licensed with an installed capacity of 1.2 million tons. However, the prominent processed items are fruit pulps and juice, fruit based ready–to-serve beverages, canned fruits and vegetables, jams, squashes, pickles, chutneys, dehydrated vegetables, etc. More recently, products like frozen pulps and vegetables, frozen dried fruits and vegetables, fruit juice concentrates and vegetables, canned mushroom and mushroom products have also been taken up for manufacture by the industry. Presently new methods of processing in India include high voltage pulse techniques, photodynamic inactivation, microwave processing- heating, high pressure treatment, ionizing radiation, heating of electric resistance effect and induction are followed in addition to popular improved traditional methods. India has remained consistently a net exporter of agriculture products, touching Rs 2.7 lakh crores exports and imports at Rs 1.37 lakh crores in 2018-19 as per economic survey 2019-20. India's exports of processed food were

Rs. 31, 111.90 Crores in 2018-19, which includes the share of products like mango pulp (Rs. 657.67 Crores/ 93.97 USD Millions), processed vegetables (Rs. 2473.99 Crores/ 354.75 USD Millions), cucumber and gherkins prepared & preserved (Rs. 1436.08 Crores/ 205.84 USD Millions), processed fruits, juices & nuts (Rs.2804.97 Crores/ 402.52 USD Millions). On the other hand, India spent for importing Rs.12,550.16 crores for 10.10 lakh tons of fresh vegetables and fruits; Rs.937.18 crores for 68,636 tons of processed vegetables and fruits & (Rs.136.46 crores for 6243 tons of flowers from different countries of world during 2018-19 (APEDA).

Crop biofortification varieties: After attaining food security, India's major initiative was to breed food crops with better nutritional values. The efforts helped to develop about 85 biofortified varieties of different crops by 2021. Crop varieties have been made nutritionally rich and superior through breeding methods. This is the most sustainable and cost-effective approach to tackle malnutrition among masses. Newly released biofortified crop varieties are rich in protein, zinc, iron, vitamin A, vitamin C, lysine (for protein synthesis) and tryptophan (for protein synthesis). The biofortified varieties are 1.5 to 3.0 times more nutritious than the traditional varieties. The biofortified varieties of different crops are given below. Rice variety CR Dhan is having 10.3 % protein, DRR Dhan 45 having 22.6 ppm Zinc, DRR Dhan 48 having 24 ppm Zinc, DRR Dhan 49 having 25.2 ppm Zinc, CR DHAN 311 ((Mukul) having 20.1 ppm Zinc and protein 10.1 %, CR Dhan 315 having 24.9 ppm Zinc; Wheat variety WB 02 is having 40 ppm Iron & 42 ppm Zinc, HD 3298 having 12.1 % protein & 43.1 ppm Iron, DBW 303 and DDW 48, each variety having protein 12.1% ; Maize pusa hqpm 7 hybrid is having 7.10 ppm provitamin A, lysine 4.19% and tryptophan 0.93 %, maize lqmh 201 hybrid having 3.03% lysine & tryptophan 0.73 % ; Finger millet varieties CFMV 1 and 2 are rich in calcium, iron and zinc. The CFMV 1 variety of finger millet is having 428 mg/100 g calcium, 58 ppm iron & 44 ppm zinc; Mustard variety, Pusa Double Zero Mustard 31 variety is low in erucic acid (0.76 % in oil) and glucosinolates (29.41 ppm in seed meal) in comparison to >40.0 % erucic acid (increases blood cholesterol) and >120.0 ppm glycosylates in popular varieties, Ground nut varieties Girnar 4 and 5 are rich in increased Oleic Acid (78.4 %); Yam's Shri Neelima and DA 340 varieties are enriched with Anthocyanin; Pomegranate variety Solapur Lal fruits rich in iron, zinc, vitamin-C and anthocyanin (Biofortified Varieties: Sustainable Way to Alleviate Malnutrition, ICAR,2020).

Nano Liquid Fertilizers: Revolutionizing Indian agriculture: To improve the yield of agricultural crops, major elements like nitrogen and phosphorous are supplied through fertilisers. However, the percentage of the fertilizers taken up by the plants is much less than the quantity of fertilizer applied. Therefore, the farmers are forced to apply more fertilizers for the need of the plants. According to some preliminary estimates, the current subsidised rate for a 50-kilogram bag of DAP costs around Rs 1,350-1,400 to the farmer, but a 500 ml bottle of Nano DAP, which may cost around Rs. 600 (Kumar *et al.*, 2023). Every year, there is a demand for 10-12.5 million tonnes of DAP, but only 4-5 million tonnes are produced in India; the remaining demand is met through imports. The difference between India's domestic urea production is approximately 26 million tonnes and the country's annual need of approximately 35 million tonnes which is met through imports.

The Indian Farmers and Fertilizer Cooperative (IFFCO) has developed indigenously through proprietary technology the liquid formulations of nano Urea and nano DAP to replace the conventional forms of these fertilisers at their Nano Biotechnology Research Centre (NBRC) in Kalol, Gujarat. The size of one IFFCO Nano Urea particle is about 30 nanometres (1 nm is one-billionth of a meter), and when compared to the conventional/ granular urea, it has about 10,000 times more surface area to volume size while the size of IFFCO nano DAP is less than 100 nm. These liquid fertilizers are sprayed directly on the leaves, thus, making systemic absorption easy for the plant. The fertilizer dose of 500 ml/acre of these nano fertilizers applied twice is ideal for the crop plant.

The shelf life of nano urea is one year, and farmers don't need to worry about its "caking" when it comes into touch with moisture because it has such a long shelf life. The efficiency of regular urea is approximately 25%, whereas the efficiency of liquid nano urea can reach as high as 85%–90%. It is asserted that a nano urea spray bottle with a capacity of 500 millilitres can fulfil the same function as a complete bag of urea weighing 45 kilogrammes (Tomar *et al.*, 2023). A 45-kg bag of urea costs around Rs 3,000 and is sold to farmers at a subsidised cost of Rs 242. But 500 millilitres of 'nano urea' costs Rs 240. Nano urea liquid is already included in recommended fertilizer doses based on multi-location and multi-crop trials undertaken in National Agriculture Research System (NARS), ICAR Research Institutes, State Agriculture Universities and Krishi Vigyan Kendra. The Union Minister of State for Chemicals and Fertilizers, BhagwanthKhuba, stated that IFFCO has set up a Nano DAP plant at its Kalol unit, Gujarat, with a production capacity of 2 lakh bottles of 500 ml per day. The plant will have a production capacity of 2 lakh bottles of 500 ml per day. In the case of nano urea, it is anticipated that production would reach around 440 million bottles of 500 millilitres each by the end of the fiscal year 2025. This quantity of nano will be equivalent to approximately 20 million tonnes of urea.

Livestock and Poultry: The contribution of livestock to agriculture and allied sectors in terms of Gross Value addition (GVA) is 28.36%, and stands at 5.21 % of total GVA during 2019-20 (Source: Annual Report 2020-21, Department of Animal Husbandry and Dairying, Government of India). It is estimated that 30% of the total protein requirement for the human population is derived from animal husbandry. The livestock and poultry provide food items such as milk, meat and eggs for human consumption. Besides, bullock (castrated male domestic bovine animal) is the backbone of Indian agriculture. Despite a lot of advancements in the use of mechanical power in Indian agricultural operations, the Indian farmers especially in rural areas still depend upon bullocks for various agricultural operations. Dung and other animal wastes serve as very good farm yard manure and the value of it is worth several crores of rupees. In addition, dung is also used as fuel. Livestock is a source of subsidiary income for many families in India, especially the resource poor who maintain few heads of livestock. The farmers in India maintain mixed farming system i.e., a combination of crop and livestock where the output of one enterprise becomes the input of another enterprise thereby realize the resource efficiency. Livestock sector has made phenomenal growth during the last eight decades with many-fold increase in production and productivity of livestock, bringing India towards self-reliance in animal products as well as increasing export potential of this sector. India has the largest bovine population of the world. Total livestock population is 536.76 million in India with 36% constituted by cattle and 20% by buffaloes during 2019-20. The total livestock population (in million) in major eastern Indian states are Assam (18.09); Bihar (36.54); Chhattisgarh (15.87), Jharkhand (23.61); Odisha (18.17) and West Bengal (37.48). Indigenous breeds of cattle were improved through field progeny testing and selection; conservation efforts increased their population. Desi cattle breeds are important because of their resistance to tropical diseases, low maintenance cost besides providing A 2 milk. Indian scientists have developed technological innovations and solutions for efficient and sustainable buffalo production systems. Conservation and propagation of most precious high yielding elite buffalo germplasm is a well-recognized and widely praised by the scientific fraternity. Number of elite bull mothers of Murrah and Nili-Ravi breeds of buffalo have been increased in the country. India developed first cloned buffalo (Asexual mean of reproduction to produce genetically identical copies of any animal) in the world in 2009. One of the biggest advantages of this technique is that the calf of desired sex can be obtained. Goat is considered as small holders' cow for means of livelihood security, poverty alleviation and employment generation. Scientists emphasized on more lambs per lambing, and successfully developed and propagated Avishaan sheep with ability to produce more offspring (2-4 lambs/lambing) for higher mutton production to enhance income. The efforts helped to obtain higher body weights at slaughter age besides overall improvement in sheep production system. The sheep husbandry sector contributes nearly 8% of total meat production, and employs nearly 6 million people in the country. India stands second in wool production in the world.

Poultry farming is one of the fastest growing sectors of agriculture transformed from a traditional backyard activity to Agri based industry. The growth rate is 8.51% in egg and 7.52% in broiler production. Around 30 million farmers are engaged in backyard poultry farming as per 19th Livestock Census. India is the third largest egg producer in the world. New poultry varieties have the potential to increase farmers' income by 2.5 times. Aseel chicken is most popular Indian game bird (used for fighting) and seen in Odisha, Andhra Pradesh and Chhattisgarh. Ankaleshwar, a hardy desi breed, reared mainly by tribals for meat and egg production under backyard system. Kadaknath, a black bird developed for nutritionally rich black meat. At present Vanraja breed is dominating backyard poultry. The consumption pattern of poultry products is rising due to its advantages in terms of low cost and high nutritive value. Poultry meat and eggs are the cheapest animal protein available and consumed globally across diverse cultures, traditions and religions. Traditional small scale/ rural poultry system continues to play a crucial role in sustaining livelihoods, supplying poultry products in rural areas, and providing important support to women farmers. The beautiful poultry bird of Indian origin was first domesticated during Indus valley civilization around 2500-2100 BC. While moving to other regions of the world, it has contributed to the evolution of various breeds of domestic chicken across the globe. India has four of the 34 biodiversity hot-spots in the world. Poultry sector in India is valued at about Rs. 80,000 crore (2015-16) broadly divided into two sub-sectors, one with a highly organized commercial sector with about 80% of the total market share (say, Rs. 64,000 crore) and the other being unorganized (rural/ backyard) with about 20% of the total market share of Rs. 16,000 Crore. The unorganized sector also referred to as backyard poultry which plays a key role in supplementary income generation and family nutrition to the poorest of the poor. Discussions with various stakeholders reveal that poultry sector, especially commercial poultry sector, is flourishing in certain pockets, where amenable environment exists, along with backward and forward linkages while the unorganized sector is very dispersed and micro-fragmented (Source: National Action Plan for Egg & Poultry-2022 For Doubling Farmers' Income by 2022, GOI). It is estimated that with a poultry population of 851.80 million, small and medium farmers are mostly engaged in contract farming systems under larger integrators and more than 30 million farmers engaged in backyard poultry during 2019-20. The total poultry population (in million) in those eastern states are Assam (46.71); Bihar (16.52); Chhattisgarh (18.71); Jharkhand (24.83); Odisha (27.43) and West Bengal (77.32) during 2019-20.

Vaccines and diagnostics developed by the scientists have made improvement of health and enhanced livestock productivity. The eradication of rinderpest reduced import dependence of livestock products and increased export of meat. Vaccines were also developed against important small ruminant diseases such as PPR (Peste des petits ruminants), goat pox and sheep pox.

The total Livestock & Poultry population (in million) during 20th livestock census 2019 in major eastern states and comparison with India are given in Table 4.

Table 4. Total Livestock & Poultry population (million) during 20th livestock census 2019 in major eastern states and comparison with India (2019-20).

State	Cattle	Buffalo	Sheep	Goat	Pig	Total livestock population including others	Total poultry population
Bihar	15.39	07.71	00.21	12.82	00.34	36.54	16.52
Chhattisgarh	09.98	01.17	00.18	04.00	00.52	15.87	18.71
Jharkhand	11.22	01.35	00.06	09.12	01.27	23.61	24.83
Odisha	09.90	00.45	01.27	06.39	00.13	18.17	27.43
West Bengal	19.07	00.63	00.95	16.27	00.54	37.48	77.32
India	193.46	109.85	74.26	148.88	09.05	536.76	851..80

Source: Annual Report 2020-21, Department of Animal Husbandry and Dairying, Government of India

Egg: Egg is a wholesome, nutritious food with high nutrient density because, in proportion to their calorie count, they provide 12% of the Daily Value for protein and a wide variety of other nutrients like vitamins, essential amino acids and minerals such as vitamin A, B6, B12, folate, choline, iron, phosphorus, selenium, and zinc etc. along with various other important ingredients so crucial for growth and good health. Protein in nutrition is one of the most important health indices that affect children's growth and development. India is one of the leading countries in poultry production ranking 3rd position in egg production. During 2019-20, India produced 114383.1 million eggs from 851.80 million poultry population, out of which 63% contributed from commercial poultry and 37% from backyard poultry. The present per capita per annum availability of egg is 86, which is far below the ICMR recommendation of 180 eggs.

The production of eggs (in million) in major eastern Indian states during 2019-20 are Assam (514.9) from total poultry population of 46.71 million; Bihar (2740.8) from total poultry population of 16.52million; Chhattisgarh (2028.9) from total poultry population of 18.71million; Jharkhand (692.8) from total poultry population of 24.83 million; Odisha (2381.4) from total poultry population of 27.43 million; and West Bengal (9735.0) from total poultry population of 77.32 million. Improved fowl contributes 87.33% and 11.52% from desi fowl with respect to total production of eggs. Besides desi duck and improved duck contributes 0.89% and 0.26% respectively with reference to total egg production in India.

Among eastern Indian states, West Bengal contributed 8.5% (4th position in egg production among all states), Odisha 2.0% and Chhattisgarh 1.8% of India's total egg production during 2019-20. The production of eggs (in million) in major eastern Indian states and comparison with India during 2019-20 are given in Table 5.

Meat: The total meat production in India was 8.6 million tonnes during 2019-20. The poultry meat production was 4.34 million tonnes (6th largest poultry meat production in world), contributing 50 % of total meat production in India. The top 5 meat producing states are UP (13.5% of total meat production in the country), Maharashtra (13.2%), West Bengal from eastern India (10.5%), Andhra Pradesh (9.88%) and Telangana (9.86%). These five states together contribute 56.94 % of total meat production in the country. The production of meat (million tonnes) in major eastern Indian states are Assam (0.053); Bihar (0.384); Chhattisgarh (0.066) Jharkhand (0.067); Odisha (0.205) and West Bengal (0.903) during 2019-20. The production of meat (million tonnes) in major eastern Indian states are Assam (0.053); Bihar (0.384); Chhattisgarh (0.066); Jharkhand (0.067); Odisha (0.205) and West Bengal (0.903) during 2019-20. The per-capita per annum availability of meat is 6.45 kg against the recommendation (ICMR) of 11 kg meat per annum.

Milk: In India the livestock sector plays an important role for livelihood support to more than 8 crores of households engaged in dairying. India is the largest producer of milk in the world. In India, the dairy sector accounts for about 5.21 % of total GVA of India (at current prices) and 28.36% of the GVA of Agriculture and allied sectors. Total milk production in India was 210 million tonnes (largest producer of milk in the world) in the year 2020-2021. The top 5 milk producing states are Uttar Pradesh (16% of total production in India); Rajasthan (12.8%); Madhya Pradesh (8.6 %); Gujarat (7.7%) and Andhra Pradesh (7.69%). These 5 states together contribute 52.79 % of total milk production of India. The per capita milk availability in India is nearly 406 g per day during 2019-20. Goat milk contributes to 3% of the total milk production across the country. The contribution of exotic cows in total milk production is 3%. The average yield rate (kg/day) of milk from different species in India during 2019-20 are exotic cows (11.88), cross bred cows (8.09), indigenous cows (3.9), non-descriptive cows (2.57), buffalo (6.43), non-descriptive buffalo (4.51) and goat (0.44). The value of milk output is more than Rs 8.38 crores during 2019-2020 at current prices which is more than output value of wheat and paddy together. But the average productivity of milk is 5.10 kg per animal per day during 2018-19 (Annual Report 2020-21, Department of Animal Husbandry and Dairying, Government of India).

The total production of milk (in million tonnes) in major eastern Indian states are Assam (0.92); Bihar (10.48); Chhattisgarh (1.67); Jharkhand (2.32); Odisha (2.37) and West Bengal (5.86).

The total production of milk (million tonnes), egg (million) and meat (million tonnes) in major eastern Indian states and comparison with India during 2019-20 are given in Table 5.

Table 5. Total production of milk (million tonnes), egg (million) and meat (million tonnes) in major eastern Indian states and comparison with India during 2019-20.

State	Milk	Egg	Meat
Assam	0.92	514.9	0.053
Bihar	10.48	2740.8	0.384
Chhattisgarh	1.67	2028.9	0.066
Jharkhand	2.32	692.8	0.067
Odisha	2.37	2381.4	0.205
West Bengal	5.86	9735.0	0.903
India	198.44	114383.1	8.599

Source: Hand Book of Statistics, Indian States, 2020-21, Reserve Bank of India.

Fisheries and Aquaculture: Fisheries are solely related to catching wild fish or raising and harvesting fish through aquaculture. Aquaculture is also called "fish farming," and it involves the natural or controlled cultivation of shellfish, fish, and seaweed in freshwater and marine environments. In India, the growing consumption of fish, owing to numerous health benefits in enhancing digestion, improving skin health, boosting metabolism, etc., is primarily driving the aquaculture market. Fisheries is an important sector in India, providing employment to millions of people. Presently, Fisheries and aquaculture are fast-growing sectors of India's economy. The market growth of the fisheries sector is increasing due to shifting of consumer preferences from high-calorie meat products towards a protein-rich diet, including fish, shrimps etc.

India is the second largest aquaculture producer of the world. Fisheries sector has been showing a steady growth in the total gross value- added to about 7.28 % share of agriculture GDP. It is providing livelihood to more than 25 million people in India, who are directly engaged in fisheries and aquaculture. India is the third largest in marine fisheries global production. Besides, the sector supports food production, providing nutritional security to the food basket, contributing to the agricultural exports. India has rich and diverse fisheries resources from deep sea to lakes, ponds, rivers etc. Marine resources are spread along the country's vast coastline. The country has a long coastline of about 8118 km and 2,02 million square km of EEZ (Exclusive Economic Zone). Fish production in India from 0.75 million tonnes in 1950-51 increased to 14.164 million tonnes in 2019-20. The total fish production during 2019-20 includes 10.437 million tonnes from inland fisheries (73.7%) and 3.727 million tonnes from marine fisheries (26.3 %). Major marine fishes include sardine and mackerel whereas the freshwater catches are dominated by carps. The different species in inland fisheries are major carp (catla, rohu, mrigal), minor carp, exotic carp, Murrells (Ophiocephalus spp.), cat fishes (Wallago Attu, pangasius) and other freshwater fishes. The top 3 fish producing states are Andhra Pradesh (29.4%), West Bengal (12.58%) & Odisha (4.76%). These 3 states together contribute 54.56 % of the total fish production of India. The production of fish (in million tonnes) in major eastern Indian states are Assam (0.37); Bihar (0.64; Chhattisgarh (0.57); Jharkhand (0.02); Odisha (0.67) and West Bengal (1.78) during 2019-20.The total fish production (in million tonnes) in major eastern Indian states and comparison with India are given in Table 3. A genetically improved rohu called 'Jayanti Rohu' with 17% higher growth realization per generation was developed through systematic selective breeding and being cultivated commercially. Species diversification of freshwater aquaculture for over two dozen of important fish species such as carps, catfishes, other miscellaneous species and freshwater prawns have been successful including packages of practices of their breeding and seed production. Marine cage-culture has been used for farming high value fish species and providing technical support to install cages along the Indian coasts. Technology of composite fish culture (in which compatible and non-competing fishes are cultured simultaneously through the utilization of different feeding zones) achieved a five-fold increase in the national productivity of pond-culture carp from 0.6 t ha^{-1} yr^{-1} to 3.5 t ha^{-1} yr^{-1}. The export of marine products stood at 1.29 million tonnes and valued Rs 46,662.85 crores during 2019-20. The successful demonstration of cage culture technology of seabass, cobia, pompano and groupers in the open sea has opened-up enormous scope for utilization of vast coastal resources for the production of high-value species.

Similarly, the adoption of large-scale cage aquaculture in medium and large reservoirs needs special attention. Besides conventional pond-based farming, there has been increasing interest in the adoption of new technologies like recirculatory aquaculture systems for high-value species, biofloc technology, etc., which provide increasing scope for entrepreneurship development. Present thrust in the fisheries sector is on intensive aquaculture in ponds and tanks through integrated fish farming, carp polyculture, fresh water prawn culture, running water fish culture and development of riverine fisheries (Hand Book on Fisheries Statistics, 2020, Department of Fisheries, GOI). Production of fish (in million tonnes) in major eastern Indian states and comparison with India during 2019-20 are given in Table 6.

Table 6. Production of fish (in million t) in major eastern Indian states and comparison with India during 2019-20.

State	Fish production in million t		
	Inland	Marine	Total
Assam	0.373	0	0.373
Bihar	0.641	0	0.641
Chhattisgarh	0.572	0	0.572
Jharkhand	0.022	0	0.022
Odisha	0.660	0.015	0.675
West Bengal	1.619	0.163	1,782
India	10.437	3.727	14.164

Source: Hand book on Fisheries, 2020, Department of Fisheries, Government of India.

Per capita production of major food commodities: The per capita production of major food commodities, except pulses, followed modest to sharp increase during a period of six decades. Per capita production of cereals in the last five decades increased by 80 per cent and oilseed production by 60 per cent. But the per capita production of pulses during the 2011–20 period came down to 15.7 kg from 23.3 kg during the decade of 1960s. Per capita production of fruits, vegetables, milk, fish and meat tripled between the 1960s and the decade after 2011, while egg production recorded an eight time increase over the same period. These changes show that there was a clear shift in the composition of food output in the country in favour of horticultural and livestock commodities. In India the growth in the quantity of food produced was followed by changes in the content of the food. Food production and the basket of food show significant change over time, especially after 1981. The share of livestock products in the total value of agri-food output increased from 17.6 per cent in 1980-81 to 27.4 per cent in 2001. This increased further to 36.9 per cent in 2020-21. Similarly, it happened in the case of fruits and vegetables. Their contribution in the crop sector doubled twice in the last 70 years and the current share is around 30 per cent. The divided data on per capita production of various food commodities during various decades since 1960-61 are given in Table 7.

Table 7. Per capita production of major food commodities since 1960s (kg/year).

Item	1961-1970	1971-1980	1981-1990	1991-2000	2001-2010	2011-2020
Cereal	121.0	131.5	151.8	169.0	173.1	173.1
Pulses	23.3	18.4	16.3	14.6	12.3	15.7
Oilseeds	15.3	15.	16.8	23.2	21.6	24.1
Sugar	24.1	24.5	26.0	30.3	27.9	29.5
Fruits	23.2	31.8	33.1	41.2	49.7	69.7
Vegetables	47.6	68.9	72.4	77.3	97.7	133.5
Eggs	0.4	0.7	1.0S	1.4	2.1	3.3
Meat&Fish	4.3	4.6	5.0	6.9	8.3	13.3
Milk	43.2	42.1	55.6	70.8	88.0	121.2

Source: Ramesh Chand and Jaspal Singh, Niti Ayog, From Green Revolution to Amrit kal, July, 202 3, pp 9.

Natural Resources Management: India occupies 2.4% of global land mass and 4 %of water resources for which there is a challenge to provide livelihood support to the Indian population (18 % of global population). The land, water and forest vegetation are the basic natural resources for agricultural growth and development of a country. Land area is shrinking day by day due to population explosion, urbanization, industrialization, construction of roads and buildings and mining. Land degradation refers to loss in biological or economic productivity resulting from land uses or a process or combination of processes including climatic variations, human activities and habitation pattern. The processes contributing to land degradation include soil erosion caused by wind and/or water, deterioration of physical, chemical, biological and economic properties of soil and long-term loss of natural vegetation. In India, 37% of total geographical area is degraded i.e., ~120 million ha is under different types of land degradation (NAAS 2010). Mining results in removal of all plant cover on the land surface and productive top soil and destroys habitat for flora and fauna. Most of the mining sites are located in reserve forests and ecologically fragile zones. Unscientific mining results in immense damage to fragile ecosystem in form of soil erosion, depletion and pollution of water resources, land degradation, disruption of communication system and floods. Waterlogging and salinity in different irrigated command areas, decline in ground water level, sea water intrusion in coastal areas, drying of wet lands and low flow in natural streams have been seen due to unplanned development and management of water resources. Agricultural scientists have developed a number of models in different agroecological zones of the country to address the above two challenges and increase production of different crops.

Global climatic changes are already adversely affecting crops, soils, water, biodiversity, livestock and fisheries. Agriculture contributes to global warming primarily through the emission of methane and nitrous oxide. With systematic indigenous research, the methane emission estimate was rationalized to 3.3 million tonnes per year and a detailed inventory of greenhouse gas emission from Indian agriculture was prepared (Pathak 2015). The estimate provided strength to the rice producing developing countries for global negotiations on climate change. Info Crop, a crop simulation model for tropical environments (Andaman and Nicobar, Lakshadweep, Western Malabar region and south Assam), has been developed to assess the impacts of climate change and optimize the input use for higher efficiency and lower emission. Several mitigation and adaptation technologies for climate resilient agriculture has been developed, which will help the country to attain net zero emission and carbon neutrality to fulfil India's commitment to the United Nations. Under NICRA (National Innovation on Climate Resilient Agriculture) project, a Vulnerability Atlas of the country has been prepared and 446 climate-friendly villages in 151 clusters have been developed, which will be upscale up to 300 climate resilient villages by 2024. Technical backstopping has been provided to the states for implementation of the District Agricultural Contingency Plans, which have been prepared for more than 600 districts of the country. Facilities such as high through-put phenotyping platforms(platform that can image at least hundreds of plants daily) , free air temperature elevation (which artificially induce increased canopy temperature in field conditions),carbon dioxide and temperature gradient tunnels (A facility to study the interactive effects of carbon dioxide and Temperature), rainout shelters(to protect a certain area of land against receiving precipitations so that an experimentally controlled drought stress can be imposed on that area), animal calorimeter(Methods for measuring heat production and energy retention), shipping vessel(verifying carbon dioxide emission), flux towers(can measure the exchanges of carbon dioxide, water vapor, and energy between the biosphere and atmosphere) and satellite data receiving station for climate change research have been established at various ICAR institutes. (SK Chaudhari, S Bhaskar, A. Islam and BP Bhatt, ICAR,2022). The efficiency of the surface irrigation system can be improved from about 35-40% to around 50-60% and that of groundwater from about 65-70% to 72-75% (Planning Commission 2009).The National Water

Mission under the National Action Plan for Climate Change has set the target to improve the efficiency of water use by at least 20%. The climate change impact would further affect the agricultural production systems in India. The recent study of World Bank revealed that India would lose 2.8% of its GDP by 2050 on account of climate change impacts causing significant reduction in living standards (Mani *et al.* 2018). The major challenges for sustainable agriculture in the coming years include increase in rainfall, high inter annual variability, intense and frequent heat waves, increase in temperature (1.5 to 4.0 degree C) and rise in sea level (IPCC 2021). These stresses may hit the small farmers the most. Production systems of the small holders are risk prone and challenging. However, diversified agriculture could help sustain the natural resources and increase the income of small farms.

The priority of agriculture developments is conservation and restoration of natural resources. Agriculture scientists have developed land resources inventories for effective and sustainable utilization of natural resources. The important spatial tools for resource use planning include soil resource maps of the country (1:1 million scale), states (1:250,000 scale) and 55 districts (1:50,000 scale); soil degradation map of the country (1:4.4 million scale) and state soil erosion maps (1:250,000 scale). Geo-referenced soil fertility maps for major, micro and secondary plant nutrients have also been developed for a large number of districts of the country to facilitate optimum nutrient use at farm level and prepare fertilizer plans at macro level. Data on wasteland are used for implementation of reclamation and conservation measures. Gypsum (good source of both calcium and sulphur) - based technology has been developed to help reclamation of more than 1.3 million ha degraded land in India. Management of sodic soils can be made after measuring the soil salinity in the field. Large scale preparation and distribution of soil health cards in the country are used by the farmers to assess the current status of soil health. Model watersheds have been developed for various regions of the country to address land degradation, enhance ground water availability besides restricting land degradation and improving ecosystem services. A rubber dam technology for water harvesting in micro watersheds and micro-rainwater harvesting structure 'Jalkund' for north-east region with storage capacity of 30,000 litres have been developed (Sanjay Kumar Ray etal, Overview of natural resource conservation with respect to North-Eastern hilly agro-eco system, NRC Book Chapter, Jan, 2016).

India ranks first in rainfed agriculture globally in both area and the value of produce. Rainfed agriculture occupies about 51 percent of country's net sown area and contribute substantially toward food grain production including 44%

of rice, 87% of coarse cereals (sorghum, pearl millet, maize),90% of minor millets, 40 % of total food production and 85% of food legumes, 72% of oilseeds and 65% of cotton. Rainfed agriculture is complex, highly diverse and risk prone. It is characterized by low levels of productivity and input usage coupled with vagaries of monsoon due to climate change; resulting in wide variation and instability in crop yields. In view of the growing demand for food grains in the country, there is a need to develop and enhance the productivity of rainfed areas. Rainfed areas have tremendous potential to contribute a larger share in food production and faster agricultural growth compared to the irrigated areas which have reached a plateau. Climatic risks like droughts, unseasonal high intensity rainfall events and floods, poor water and nutrient retention capacity of soil and low soil organic matter impact highly vulnerable, rainfed agriculture requiring a different outlook and strategy. All India coordinated Research Project (AICRP) for Dryland Agriculture and Central Research Institute for Dryland Agriculture (CRIDA) have developed a number of dry land technologies/practices such as in situ and ex situ rain water harvesting and conservation, alternate and efficient cropping systems, integrated nutrient management, drought tolerant varieties etc and are integrated in to package of practices in the states and have been disseminated through KVKs/SAUs. Micro irrigation and drip fertigation technologies developed for various crops and cropping systems save up to 60% irrigation water & 40% fertilizers. Dryland technologies have been integrated in to National schemes/programme such as the Mahatma Gandhi National Rural Employment Guarantee Act (MGNREGA), National Mission on Sustainable Agriculture (NMSA), Rastriya Krishi Vikas Yojana (RKVY), Pradhan Mantri Krishi Shinchayee Yojana (PMKSY), National Rural Livelihoods Mission (NRLM) etc. Rainfed Area Development (RAD) is one of the components of National Mission for Sustainable Agriculture (NMSA). The NMSA is one of the 8(eight) Mission outlined under National Action Plan on Climate Change (NAPCC).

Integrated Farming System (IFS)is a mix of farm enterprises such as crop, livestock, aquaculture, poultry, sericulture and agroforestry to achieve economic and sustained agricultural production through efficient utilization of resources. Cluster based development to promote IFS is culturally and socially acceptable to the local community. Single cropping system is not recommended under National Mission for Sustainable Agriculture (NMSA). It is proposed for diversification with additional income generating activities. More than 65 Integrated Farming System (IFS) models have been developed for different agroclimatic zones in India for augmenting farm productivity and income of small and marginal farmers. This can help in achieving doubling of farmers income. Integration of less land requiring activities such as mushroom, apiary, bio-gas and other agriculturally related off-farm activities hold the key to achieve this vision. Agroforestry combines agricultural and forestry technologies to create more diverse, productive, profitable, healthy, and sustainable land-use systems. Based on promising tree species for a particular agroclimatic zone, the agroforestry system is characterized for different agroclimatic zones. Agroforestry provides a unique opportunity to combine the twin objectives of climate change adaptation and mitigation. It has the ability to enhance the resilience of the system to cope the adverse impacts of climate change. Agroforestry also have indirect benefit through C sequestration of elemental carbon in the trees and it helps to decrease pressure on natural forests, which are the largest sinks of terrestrial C. Another indirect avenue of C sequestration is through the use of agroforestry technologies for soil conservation, which could enhance C storage in trees and soils. (Ram Newaj *et al.*, 2016). Agroforestry has evolved as an assured land use system against crop failure as it plays a role in reducing vulnerability, increasing resilience of farming systems and buffering households against climate related risks. The scientists have developed models of agroforestry to promote Agri -Horti, Agri -Silviculture and Silviculture -pastoral systems for raising farmers' income.

Organic farming is a production system, which avoids or largely excludes the use of synthetic fertilizers, pesticides, growth regulators and livestock feed additives (USDA, 1980). This system largely depends on crop rotations, crop residues, animal manures, green manures, off-farm organic wastes, mechanical cultivation, mineral bearing rocks and aspects of biological control to maintain soil productivity, supply plant nutrients and to control insects, pathogens and weeds (Sharma, 2002). A total of 68 organic packages have been developed for promoting organic farming and integrated in central sector schemes of Paramparagata Krishi Vikas Yojana (PKVY) and Mission on Organic Value Chain Development (MOVCD) for NEH Region. Organic certification system is a quality assurance initiative, intended to assure quality, prevent fraud and promote commerce, based on set of standards and ethics. It is a process certification for producers of organic food and other organic plant products. Organic farming certification in India is done by National Programme for Organic Production (NPOP)- APEDA (Agricultural & Processed Food Products Export Development Authority). Besides certification is also done through Participatory Guarantee System (PGS). In India organic farming area (in million hectares) under NPOP certification was 2.65 and PGS certification 1.51 during 2020-21.

Presently major challenges in Indian Agriculture are widespread land degradation, spatial and temporal variability of monsoon rainfall resulting in frequent drought and floods, deteriorating soil health, low nutrient and water use efficiency and low water productivity. These challenges are very much detrimental to the small farmers due to less capacity to purchase the inputs. Hence agricultural scientists have designed and developed new cropping system for irrigated and rainfed areas, integrated farming systems for different strata of farmers, micro level land use planning, bio-engineering measures for soil and water conservation, reclamation and amelioration technologies for salt affected, waterlogged and acid soils, integrated nutrient management, resource conservation technologies, conjunctive use of rain, surface and groundwater resources, efficient and energy saving irrigation methods, multiple use of water and participatory watershed management including agroforestry interventions for sustainable natural resource management. These technologies along with application of geo-informatics, information and communication technology as well as climate smart agricultural practices may go a long way in managing natural resources. Organic farming area (in million hectares) under NPOP certification and PGS certification in major eastern area states during 2020-21 as compared to India are given in the Table 8.

Table 8. Organic farming area (in million hectares) under NPOP certification and PGS certification in major eastern area states during 2020-21 as compared to India.

State	**Organic farming area (in million hectares)**	
NPOP (National Programme for Organic Production)	**PGS Guarantee Certification**	**(Participatory System)**
Assam	0.018	0.004
Bihar	0.029	0.024
Chhattisgarh	0.023	0.109
Jharkhand	0.053	0.008
Odisha	0.092	0.044
West Bengal	0.006	0.002
India	2.657	1.151

Source: Agricultural Statistics at a Glance 2021, Ministry of Agriculture and Farmers Welfare, Government of India.

Conservation of vegetation, wildlife and livestock and bio-diversity is of paramount importance for sustainability point of view. Modern agriculture has affected water resources through the destruction of riparian habitats within watersheds. The conversion of wild habitat to agricultural land reduces fish and wildlife through erosion and sedimentation, the effects of pesticides, removal of riparian plants and the diversion of water. The plant diversity in and around both riparian and agricultural areas should be maintained in order to support a diversity of wildlife. This diversity will improve natural ecosystems to aid in managing agricultural pests Forest cover in India is estimated to be 67.83-million-hectare accounting for 20.64 % of geographical area of the country. Per capita availability of forest land in India is 0.08 ha as against global average of 0.64 hectare. For a sound ecosystem, area under forest should be one-third of the geographical area of the country. The grazing intensity is 42 animals per hectare of land as against the threshold level of 5 per hectare.

Agricultural engineering: Agricultural mechanization, has shifted from subsistence farming to robotics (involves the creation of robots to perform tasks without further intervention) and artificial intelligence/A.I. (how systems emulate the human mind to make decisions and 'learn. A.I. revolves around enabling machines to make complex decisions autonomously) between the years 1900 and 2022. There is a linear relationship between availability of farm power and farm yield. Therefore, there is a need to increase the availability of farm power from 2.49 kW per ha (2018-19) to 4.0kW per ha by the end of 2030 to cope up with the increasing demand for food grains. Farm implements and machines were developed for improving farm mechanization and save time and labour, reduce drudgery, cut down production cost, reduce post-harvest cost. Water, energy and fertilizer management and utilization of conventional and non-conventional energy sources in agricultural production and processing activities are some of the areas, which recorded tremendous success in independent India. At present the improved manual tools and animal drawn farm equipment on individual ownership basis are best suited to small farms mechanization. For high-capacity farm machinery custom hiring mechanism is the tested model in the rural areas and becoming popular. The agricultural engineering research work started in 1950s with introduction of improved agricultural equipment and machinery. Farm mechanization has been increasing steadily over the years as evident for sale of tractors and power tillers from 2014-15 to 2020-21. In India the sale of tractor (in lakh no) was 6.81 during 2014-15, which increased to 8.99 during 2020-21. Similarly, the sale of power tiller (lakh no) was 0.460 during 2014-15, which increased to 0.540 during 2020-201. The sales of tractors (in lakh no) in major eastern India states during 2021-22 were Bihar (0.370), Chhattisgarh (0.255), Odisha (0.160), West Bengal (0.157), Jharkhand (0.108) and Assam (0.080) as against total sale of 8.42 in India (Agriculture Statistics at a Glance, 2021, GOI). The combine harvesters were introduced in Northern India at the beginning of green revolution and their number grew from 800 in 1971 to over 50,000 at present (Indian Agriculture after Independence, July 2022, ICAR, New Delhi).Water saving devices like sprinkles and drip irrigation were given high importance. Micro-irrigation is an effective scientific alternative for higher irrigation efficiency of 70-90%. The micro irrigation is applicable to wider spaced perennial horticultural fruit crops. The micro-irrigation has also been developed for field crops like wheat, sugarcane, vegetables, pulses etc. Micro irrigated areas in India during 2020-21 was 13.476 million hectares which includes 6.321 million hectares under drip irrigation and 7.155 million hectares under sprinkler irrigation. In major eastern Indian states micro irrigated areas (million hectares) during the same year were Chhattisgarh (0.362), Odisha (0.144), Bihar (0.120), West Bengal (0.103), Jharkhand ((0.043) and Assam (0.020). The farm technologies pushed agriculture into the economic edge and produced lots of commercial crops that have established new agri-businesses.

The micro irrigated areas (million hectares) through drip & sprinkler irrigation during 2020-21 in major eastern India states as compared to India are given in Table 9.

Table 9. Micro irrigated areas (in million hectares) through drip & sprinkler irrigation during 2020-21 in major eastern India states as compared to India.

State	Micro irrigated area (in million hectares)		
	Drip	Sprinkler	Total
Assam	0.004	0.016	0.020
Bihar	0.014	0.106	0.120
Chhattisgarh	0.031	0.331	0.362
Jharkhand	0.025	0.018	0.043
Odisha	0.029	0.115	0.144
West Bengal	0.010	0.093	0.103
Total	6.321	7.155	13.476

Source: Agricultural Statistics at a Glance 2021, Ministry of Agriculture and Farmers Welfare, Government of India.

After the year 2010, agricultural mechanization and post-harvest processing entered the new age of technology called precision agriculture and post-harvest processing. The future agricultural engineering challenges are to stimulate right solutions for agriculture value chain through advanced technologies using the principles of precision farming involving equipment for unmanned operations and autonomous decision support system.

Agricultural Education: Agriculture education in India first started during 1901-1905 with 6 agricultural colleges established in different parts of the country. Subsequently the Imperial Council of Agricultural Research was established on 16 July 1929 as a registered society under the Societies Registration Act, 1860 in pursuance of the report of the Royal Commission on Agriculture. Agricultural higher education was weak in 1947. The University Education Commission reported that there were only 17 agricultural colleges in 1948 in India, 12 of which had been established since 1940. Facilities for training in postgraduate research work in the agricultural sciences in 1948 were available for only 166 students.

Indian Agricultural Research Institute (IARI) originally established in 1905 at Pusa (Bihar) and shifted to New Delhi in 1936. IARI is the country's premier institute for agricultural research, education and extension. It was made Deemed University in 1956 by the UGC. Thereafter award of masters and Ph.D. degrees in various agricultural disciplines started. The first veterinary college was established at Mathura in 1947. Currently, there are 17 universities offering degrees in various courses of veterinary and allied sciences. The course on agricultural engineering leading to bachelor degree was initiated for the first time in 1942 at Allahabad Agriculture Institute. First Fisheries College was established at Mangalore in 1969.

In Independent India, the state Agricultural universities were established as per the Land Grant pattern of USA on the basis of recommendation of the first joint Indo-American Team during 1955 for upgradation of agriculture and veterinary education. U.S. government technical assistance to Indian Agricultural education has gone through three phases, (a) a modest ad hoc beginning in 1952 (b) the Agricultural Education and Research Project, 1955-1961, and (c) the subsequent Agricultural University Development Project, from 1961 to 1964 (Source: Kathleen M. Propp, 1968, The establishment of Agricultural Universities in India, A case study of role of USAID-US university Technical Assistance, published by University on Illinois college of agriculture). Since then, agricultural higher education underwent several changes as a number of state agricultural universities were established. 74 Agricultural Universities (AUs) comprising State Agricultural Universities (SAUs), Deemed Universities (DUs) and Central Agricultural Universities (CAUs) is one of the largest agricultural research, education and extension systems globally. Subsequently many private universities with state legislation established in India and impart education at the level of degree, masters and doctoral level. Presently agricultural universities in India impart education in the various disciplines of agriculture viz Agriculture, Agricultural Engineering, Forestry, Horticulture, Veterinary and Animal Husbandry, Dairy Science, Food Technology, Fisheries Science, Agribusiness Management, etc. Agriculture universities are now imparting education at the level of diploma, degree, masters and doctoral level.

Scientific Agri-education system is a prerequisite for sustainable agricultural development. Education is to develop and enhance the potential of human resources and progressively transform it to a knowledge society. Similarly,

Teacher is a person who influences others and makes a change in their lives intellectually and morally. Interpersonal relationship, human relationship and management of human resources are the core to all activities of agriculture and allied universities. The essential requirement of the quality standard of Agriculture education is the capacity to develop skilled professionals in adequate numbers. Agriculture Education is confined to agricultural research, extension and agribusiness and an obligation with full social responsibility. Agriculture Education in India is guided by National Agriculture Research and Education System (NARES) led by Indian Council of Agricultural Research (ICAR). The prescribed guidelines on academic system and academic standards are followed by agriculture, horticulture and forestry, veterinary and fisheries universities, colleges and institutions of ICAR.

New Education Policy-2020 (NEP-2020) of India has proposed several changes in the education system of India, including higher agriculture education system Under the NEP-2020, ICAR has been designated as the Professional Standards Setting Body (PSSB) of Agriculture Education. A national level Committee has been constituted by the ICAR to develop an implementation strategy to comply with various provisions of National Education Policy-2020 (NEP-2020). The proposed changes in agriculture education after implementation of NEP-2020 include transformation of Agricultural Universities/Colleges into large multidisciplinary universities, colleges, and HEIs (Higher Education Institutions) clusters/Knowledge Hubs with 3,000 or more students. The multi-disciplinarity of agricultural education shall encompass academic programmes of basic sciences, social sciences and allied disciplines of agricultural sciences. Therefore, the single stream universities under the ICAR system need to move towards multidisciplinary institutions by 2030 while continuing the focus on teaching, research and extension systems of agriculture. The implementation of this policy needs the full support of the State Governments as Agriculture is a state subject. Accordingly, the Acts and Statutes of the agriculture universities need to be amended bringing them in line with the NEP-2020 guidelines by the state governments.

Agricultural Extension: Agriculture Extension is an applied behavioural science, the knowledge of which is applied to bring farmers or people through various strategies and programmes of change by applying the latest scientific and technological innovations. Agricultural extension helped reduce the gaps in potential yields and farm realized yield. Agricultural Extension is not static. It has evolving with the times & embracing different branches of knowledge cutting across disciplines that included modern management, marketing, entrepreneurship, information & communication technologies, monitoring

& evaluation, climate change, gender studies, public policy, human resource management, health and nutritional security and many more to address emerging challenges in agricultural development (MANAGE Annual Report, 2020-21).

Indian Agriculture has a typical blend of man, machine, animal and material usage and influenced by the behavioural changes. Cultivation of a single crop in a season without use of chemical fertilisers and machines was the practice when our population was less. Intensive cultivation of crops with use of chemical fertilisers, machines started when population increased. The man-to-man interface, structured or unstructured, played a catalytic role in this evolution of agriculture.

The agricultural extension system in India evolved through phases. The extension system firstly helped the farmers in the intensive use of inputs and cultivation of dwarf varieties of rice and wheat, hybrids of maize and other coarse cereals for food security. The community development programme in 1950s, intensive area development programme, T&V programme, lab to land programme in 1970s and oilseeds & pulses promotional programme in 1980s, micro-irrigation and farm mechanizations 1990s onwards and more intensively post MGNREGA are the critical attempt by the agricultural extension system. Informal extension from farmers to farmers changed to farmers and state-owned subsidized input providing units located at the block headquarters and below and private input dealers. On the other hand, formal extension was divided into mass extension machinery owned by the state department of agriculture, and frontline extension under the umbrella structure of NARS (National Agriculture Research System) under ICAR during the 1970s. This extension system proved the most productive for many important agricultural programme-like national demonstrations programme on HYVs, technology mission on oilseeds, accelerated pulses development programme, seed hub programme, etc. Indian Council of Agricultural Research (ICAR) has established front line extension system in the form of Krishi Vigyan Kendra (KVK) at grassroot level for effective dissemination of new technologies developed by Agricultural scientists for increase of productivity and profitability in growing crops by the farmers. The first KVK established in the year 1974 in Puducherry and presently total KVKs in the country are 731. The KVK activities include on-farm testing to identify the location specificity of agricultural technologies under various farming systems, frontline demonstrations to establish the production potential of improved agricultural technologies on the farmers' fields, training of farmers and extension personnel to update their knowledge and skills. Besides, KVK works as a knowledge and resource centre of agricultural technologies for supporting farmers in improving their agricultural production and livelihoods. In view of the changing scenario of Agriculture and allied sectors, the mandated activities of KVKs are being changing from time to time to address the newer challenges in the area of climate change, secondary and speciality agriculture, conservation agriculture, nutri-sensitive agriculture, market led extension, agribusiness and special drives for disadvantaged and aspirational districts in India for agriculture.

At present agricultural extension systems in India have two broad typologies based on their domain of action- Mass Extension or Field Extension System through state agriculture extension machineries and Frontline Extension System through KVKs. The field extension system is concerned with dissemination of established agricultural technologies to the masses by state development departments and agencies and the frontline extension system concerns to testing and demonstration of new technologies in a quick succession, differently in approach and design, which otherwise take very long period to reach to the farmers, as well as capacity development of the related stakeholders. The frontline Extension system through KVKs is run in the country by ICAR institutes, State Agriculture Universities and reputed NGOs. Apart from the frontline extension, the mass extension in agriculture & allied sectors is under the governance of Department of Agriculture & Farmers Welfare (DAFW), Department of Fisheries, Department of Animal Husbandry & Dairying and the related ministries of rural development through their network of agricultural and related departments at state, district, block and village level.

KVKs have taken up various additional programme like Farmers FIRST programme (FFP), Attracting and Retaining Youth in Agriculture (ARYA), Cluster Frontline Demonstration of pulses and oilseeds, Cereal Systems Initiatives for South Asia (CSISA), National Innovations in Climate Resilient Agriculture (NICRA), Pulses Seed hubs, Mera Gaon Mera Gaurav and Awareness creation on government schemes, etc. In India 15 lakh farmers& farm women, rural youth and extension personnel are trained and technology demonstrations are conducted in more than 3 lakh farmers' field annually through KVKs.

Agricultural Research: Agricultural research started in India with the establishment of Imperial Council of Agricultural Research (1929) in Delhi, which is known today as Indian Council of Agricultural Research (ICAR). This is the apex body with its headquarters at New Delhi for coordinating, guiding and managing research and education in agriculture including animal sciences and fisheries. The Council is an autonomous organisation under the governance of Department of Agricultural Research and Education (DARE), Ministry of Agriculture and Farmers Welfare, Government of India.

Presently ICAR has 113 institutions (72 Research Institutes, 6 National Bureaus, 23 Project Directorates and Agricultural Technology Application Research Institutes, 12 National Research Centres), 82 All India Coordinated Research Projects & Network Research Projects and 74 Agricultural Universities spread across the country. This is one of the largest national agricultural research systems in the world. 731 Krishi Vigyan Kendra are functioning in the country. India has one of the largest agricultural research human resource capitals in the world with approximately 30,000 scientists and more than 100,000 technical & supporting personnel. ICAR footprints are also extended to the neighbouring countries and several international, national and regional research organizations and universities are engaged with ICAR in agricultural research and development. Additionally, private and non-Governmental organizations and farmers themselves have done significant agricultural research in their own field.

Collaborations in research with national and international agencies have major role in shaping the path for agricultural development in India. The collaboration with CIMMYT (International Maize and Wheat Improvement Centre) and IRRI (International Rice Research Institute) for exchange of dwarf wheat and rice varieties, respectively triggered the efforts for the Green Revolution making India food secure. The Indian Council of Agricultural Research (ICAR)- Consortium of International Agricultural Research Centres collaborations (CGIAR) expanded to various commodities and strategic locations for development of new crop varieties and animal breeds and also managing natural resources. Development of new breeds by harnessing the potentials of the exotic breeds, vaccines and diagnostics, quality standards, health, nutrition and hygiene standards and protocols came out from the international and national collaborations in livestock and fisheries sector.

Conclusion

The plan for reimagining Indian agriculture prioritizes enhancing farmers' income by 200%, while simultaneously targeting reductions of 25% in fertilizer use, 20% in water use, and 45% in greenhouse gas emission intensity, alongside a 50% increase in renewable energy adoption. Additionally, the rehabilitation of 26 million hectares of degraded land is planned. This comprehensive approach aims to foster sustainable agricultural practices, ensure environmental stewardship, and promote resilience in the face of climate change, thereby laying the groundwork for a more prosperous and sustainable future for Indian agriculture.

References

Agriculture Statistics at a Glance,2020, Government of India, Ministry of Agriculture and farmers Welfare, Department of Agriculture and Farmers Welfare, Directorate of Economics and Statistics

Agriculture Statistics at a Glance,2021, Government of India, Ministry of Agriculture and farmers Welfare, Department of Agriculture and Farmers Welfare, Directorate of Economics and Statistics

Amarnath Tripathi and A. R. Prasad (2009) , Journal of Emerging Knowledge on Emerging Markets ,Volume 1 Issue 1 ;Agricultural Development in India since Independence: A Study on Progress, Performance, and Determinants

Annual Report, 2020-21, Government of India, Ministry of Agriculture and farmers Welfare, Krishi Bhawan, New Delhi

Annual Report, 2020-21, Government of India, Ministry of Fisheries, Animal Husbandry and Dairying, Department of Animal Husbandry & Dairying, Krishi Bhawan, New Delhi

Annual Report, 2021-22, Government of India, Ministries of Fisheries, Animal Husbandry and Dairying, Department of Fisheries, Krishi Bhawan, New Delhi

Annual Report, 2021-22, Government of India, Ministry of Agriculture and farmers Welfare, Krishi Bhawan, New Delhi

Area and Production of Horticulture Crops for 2018-19 (Third Advance Estimates), Department Agriculture & Farmers' welfare, Government of India

Area and Production of Horticulture Crops for 2019-20 (Third Advance Estimates), Department Agriculture & Farmers' welfare, Government of India

Basic Animal Husbandry Statistics, 2019, Government of India, Ministry of Fisheries, Animal Husbandry and Dairying, Department of Animal Husbandry & Dairying, Krushi Bhawan, New Delhi

Biofortified Varieties: Sustainable Way to Alleviate Malnutrition,3rd edition, ICAR, New Delhi,2020

Chandra, D., Chandra, S., & Sharma, A. K. (2016). Review of Finger millet (Eleusine coracana (L.) Gaertn): A power house of health benefiting nutrients. Food Science and Human Wellness, 5(3), 149-155.

Dr SK Malhotra Project Director, Directorate of Knowledge Management in Agriculture, ICAR (2022), ICAR Transforming Indian Agriculture

Geervani, P., & Eggum, B. O. (1989). Nutrient composition and protein quality of minor millets. Plant Foods for Human Nutrition, 39, 201-208.

Government of India, Horticulture Statistic Division, Department of Agriculture, Cooperation & Farmers Welfare, Ministry of Agriculture & Farmers Welfare, Horticultural Statistics at a Glance 2018(Compilation: Naveen Reddy , Moreshwar Karale, Ranbir Singh, Pankaj Gilotra).

Government of India, Ministry of Agriculture and Farmers welfare, Department of Agriculture Cooperation and Farmers Welfare, Directorate of Economics and statistics, Agricultural Statistics at a Glance 2019, Chief Supervision: P.C.Bod

H. Pathak, J.P. Mishra, T. Mohapatra (2022), Indian Agriculture after Independence, Indian Council of Agricultural research, New Delhi

Hand book of Statistics on Indian Economy, 2021-22, Reserve bank of India,

Hand book of Statistics on Indian States, 2020-21, Reserve bank of India,

Hand book on Fisheries Statistics, 2020, Department of Fisheries, Ministries of Fisheries, Animal Husbandry and Dairying, , Government of India, New Delhi

https://www.agric.wa.gov.au/irrigated-crops/nutritional-aspects-quinoa

ICAR-Directorate of Mushroom Research, Solan (HP) , Annual Report , 2019

ICAR-National Institute of Abiotic Stress Management, Baramati, Maharashtra, Technical bulletin, June 2019, Quinoa an alternate food crop for water scarcity zones of India, edited by Jagdish Rane etal

ICAR Sponsored Winter School on Advancement in Potato Production Technology and its Future Prospects (November 19 to December 09, 2019) Organized by ICAR-Central Potato Research Institute Regional Station, Modipuram-250110, Meerut (UP)

Indian Agriculture Towards 2030 : Pathways for Enhancing Farmers' Income, Nutritional Security and Sustainable Food and Farm System by Niti Ayog and FAO, edited by Ramesh Chand ,Pramod Joshi & Shyam Khadka (2021), pp1-311.

Indian Farming, November, 2020, Special Issue on world Fisheries Day, ICAR, New Delhi.

Indian Horticulture, July-August, 2022, Vol 66, No4 ,ICAR, New Delhi

K. Srinivas, Upma Dubey & Navya Lalitha, Analysing the Value Chain of Quinoa: A Case Study of Quinoa, FIIB, Business Review vol 4, Issue 4, October-December,2015

Kathleen M. Propp, 1968, The establishment of Agricultural Universities in India, A case study of role of USAID-US university Technical Assistance, published by University on Illinois college of agriculture

Kuldeep Kumar etal, ICAR-Indian Institute of Soil and Water Conservation, Research Centre, Kota, Rajasthan, Quinoa: A superfood & climate resilient crop for higher income in vertisols of south-eastern Rajasthan , Harit Dhara 3(1) January-June ,2020

Kumar, N., Samota, S. R., Venkatesh, K., & Tripathi, S. C. (2023). Global trends in the use of nano-fertilizers for crop production: Advantages and constraints–A review. Soil and Tillage Research, 228, 105645

Muthamilarasan, M., Dhaka, A., Yadav, R., & Prasad, M. (2016). Exploration of millet models for developing nutrient rich graminaceous crops. Plant Science, 242, 89-97.

Presentation for International year of millets, 2023 in National Conference on Kharif Campaign, 2022 on 19th April, 2022 by Ministry of Agriculture & Farmers Welfare

Proceeding Book on Two Days National Webinar on Approaches Towards Onion Cultivation, 26-27 May, 2020, Organized by Krushi Vigyan Kendra, Khedbrahma (Sabarkantha), Sardarkrushinagar, Dantiwada Agricultural University, Gujarat.

Ramesh Chand, Transforming Agriculture for Challenges of 21st Century, 102 Annual Conference Indian Economic Association (IEA) 27-29 December 2019 Ramesh Chand, Agricultural Challenges and Policies for the 21st Century, NABARD ,2022

Ram Newaj, O.P. Chaturvedi and A.K. Handa ,Indian Journal of Agroforestry Vol 18 No,1-9 (2016), Recent development in agroforestry research and its role in climate change adaptation and mitigation

Rane J, Pradhan A, Aher L, Singh NP. (2019). ICAR-NIASM Publications, ICAR-NIASM, Baramati. Pp-12

Review of Finger millet (Eleusine coracana (L.) Gaertn): A power house of health benefiting nutrients

Saleh, A. S., Zhang, Q., Chen, J., & Shen, Q. (2013). Millet grains: nutritional quality, processing, and potential health benefits. Comprehensive reviews in food science and food safety, 12(3), 281-295.

Tajamul Rouf Shah, K. Prasad and P. Kumar (2016), Maize- A potential source of human nutrition health-A review, Congent Food and Agriculture Journal, vol2, 2016- Issue 1.

Tomar, M., Choudhary, R., & Patidar, R. (2023). Chapter-1 Nano Urea-A Bliss for Agriculture or Not. Recent Trends in Agriculture, 1

2

Development of Fisheries and Aquaculture in Eastern India

J.K. Jena[1] and P.C. Das[2]

[1]*Indian Council of Agricultural Research, Krishi Anusandhan Bhawan-II Pusa, New Delhi, India*

[2]*ICAR-Central Institute of Freshwater Aquaculture, Bhubaneswar, Odisha, India*

1. Introduction

Fish forms an integral component of the Indian food basket and it is more so for the population living in the eastern and north eastern States of India. The eastern states of India include States such as West Bengal, Bihar, Jharkhand and Odisha. As per a current estimate of 2023 (https://www.indiacensus.net), approximately a population of 31.66 crore, i.e. 22.15% of the country's total, live in these four States. Due to the suitable geographical terrain, these states have been bestowed upon a large number of riverine stretches, a large number of natural reservoirs of varied sizes, beels and floodplain wetlands, and a huge plain terrain supporting ponds and tank resources. Due to the very nature of the land and the popularity of fish among the populace, these eastern Indian states have also served as the main stage for the development of fisheries and aquaculture during the last centuries in the country. Further, many of the epoch-making technologies for the development of fisheries in the inland waters as well as aquaculture in the ponds have originated from this region.

2. Importance of Fisheries and Aquaculture in Eastern India

The four states in eastern India have 23% of the pond and tank resources of the country (DoF, 2022). Due to the homestead nature of most of these ponds, commercial grow-out farming was limited to only a few ponds until the beginning of last century but has taken center stage as commercial farming activity during the last few decades. The fish production trends in these four states since 2000 are depicted in Fig. 1. Though there was a progressive increase in production, marked change in the trend is observed mostly during the last decade. Out of the four states, West Bengal and Odisha are the two maritime states having additional fish catch from the marine sector. Since marine catch has almost remained stationary, much of the increase in production was achieved due to an increase in inland fish production.

The eastern region contributes almost 30% of the total inland production of the country. Over the years, though the inland production from this region has consistently increased, the share of contribution has been decreasing since 2006 which indicates that the growth in fish production is not at par with the overall growth trend of the country. Though marine catch has plateaued, there has been a significant increase in the fish production of the country due to a marked increase in fish production from aquaculture activity. However, a similar rise in fish production was lacking in this region, particularly in West Bengal and Odisha until recent years, mostly due to the availability of a large share of homestead pond resources with limited scope for intensification.

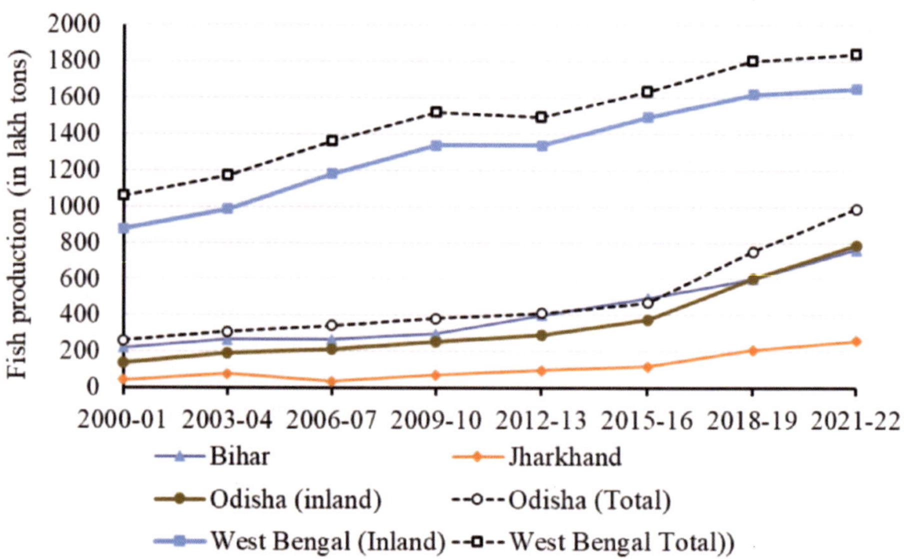

Fig. 1: Fish production in the eastern Indian states

The available resources for fish production in these four eastern states are given in Table 1.

Table 1. Fisheries Resources in the Easter Indian States

Resources	Rivers & canal (Km)	Small reservoirs (ha)	Medium & large reservoirs (ha)	Ponds & tanks (ha)	Beels & oxbow lakes (ha)	Brackish water (lakh ha)	Others (ha)
Bihar	21,300	0	26,304	93,218	950,000	-	0
Jharkhand	1,800	28,790	104,363	79,010	0	-	19,936
West Bengal	2,526	28,050	0	263,372	42,082	2.10	26,925
Odisha	24,879	34,608	165,771	136,951	180,000	4.30	0
Eastern India Total	50,505	91,448	296,438	572,551	1,172,082	6.40	46,861
Country Total	280,751	754,833	2,045,205	2,448,368	1,355,060	11.55	920, 391
Share of Eastern India to country's Total (%)	18.0	12.1	14.5	23.4	86.5	55.4	5.1

2.1 Fisheries and aquaculture in Odisha

The state of Odisha is the 4th major fish-producing maritime state after Andhra Pradesh, West Bengal, and Gujarat. Over the last 5 years, the fisheries sector has grown at an average annual growth rate of about 13% in the State. The total fish production in the State in 2020-21was 8.73 lakh tonnes, of which the share of freshwater, brackishwater, and marine sources were 5.75 lakh tonnes, 1.25 lakh tonnes, and 1.73 lakh tonnes respectively. This production has shown a spectacular increase to 10.52 lakh tonnes with an inland production of 6.90 lakh tonnes, marine production of 2.13 lakh tonnes, and brackishwater production of 1.49 lakh tonnes in 2022-23. With a 94.4% fish-consuming population, the annual per capita fish consumption of the State is reported to be 16.24 kg in 2020-21. The export of seafood from the state increased from Rs. 3,243 crores in 2019-20 to Rs. 4,807 crores in the year 2022-23.

There are 749 primary fisheries societies operating in the State. With the vast fisheries resources and fishermen population, Odisha has a huge scope for development in the fisheries sector. The State has brought the Marine Fishing Regulation Act, 1983, the Reservoir Fisheries Policy, 2012, and the Odisha Fisheries Policy,2015 which have been the principal legislations/policies for managing the fisheries resources and aquaculture development. The Odisha Fishery Policy, 2015 has mandated popularising scientific pisciculture in the state thus increasing fish production and the farmer's income. It included State plan Schemes such as: subsidy for intensive freshwater aquaculture through the excavation of new fish ponds (Matsya Pokhari Yojana); popularisation of fishery machinery/equipment/implements for intensive aquaculture; input assistance to Women SHG for pisciculture in GP tanks; promotion of intensive aquaculture through the introduction of bio-floc technology; support to private fish hatcheries for the adoption of early breeding and year-round seed production; hatchery accreditation and seed certification; input assistance to farmers for taking up fish farming in farm ponds; input assistance to develop a network of fish seed growers for early bred spawn, etc. The Policy, 2015 envisaged adding 15,000 ha of new tank/pond area in 30 districts by 2020 besides bringing 1500 ha of existing GP and Revenue ponds under scientific culture every year by leasing in for five years and with other suitable interventions. The policy also targeted to enhance the productivity of extensive farming practices from 2.13 tonnes/ha to 3.5 tonnes/ha, semi-intensive practices from 2.60 t/ha to 5 t/ha, and intensive systems. In line with the policy, several schemes have been implemented in the State during the last decade with the active financial support of the NFDB, State, and Central Governments over the years for holistic development of the fisheries and aquaculture sector (http://fisheries.odisha.gov.in). Ensuring cheap availability of rice for fishermen during the closed fishing season; providing fish vending equipment for efficient marketing of quality fish; promotion of cage and pen farming in reservoirs; replacement of wooden craft with FRP boats for marine fishers; mechanisation of traditional fishing craft; subsidy for net and boat for inland fishers; insurance coverage for fishermen; houses for fishermen under the 'Basundhara' and 'Mo Kudia' schemes; financial support to the Women SHG, etc are some of the examples of schemes implemented for the overall development of the fisheries sector. Besides, several central schemes have also been implemented for aquaculture development. It includes the Fisheries and Aquaculture Infrastructure Development Fund (FIDF); PMMSY with its sub-components of construction of new inland rearing and grow-out ponds, establishment of new freshwater finfish seed hatcheries, establishment of state-of-the-art fish feed plant, construction of the modern fish retail market, installation of cages in reservoirs, construction of fish kiosk, establishment of disease diagnostic & quality testing lab, etc.

2.2 Fisheries and aquaculture in West Bengal

Fish forms the most important component in the food basket of the people of West Bengal. The total Fish production of the state in 2021-22 was 18.43 lakh tonnes out of which inland production contributed 16.52 lakh tonnes(DoF, 2022). The fisheries sector provides livelihood to 28.34 lakhs inland and 4.25 lakh marine fishermen besides supporting the population engaged in allied activities.

Aquaculture development in West Bengal has been spearheaded over the years by the Fisheries Department. The Fisheries Department, one of the oldest in the country, was first set up in 1911, abolished in 1923 and again revived in 1942. It was renamed as the Department of Fisheries, Aquaculture and Aquatic Resources & Fishing Harbors in 2001 due to its responsibility of managing multi-directional fishery-related activities. This Department under its control has the State Fisheries Development Corporation Ltd. (SFDCL), West Bengal Fisheries Corporation Ltd. (WBFCL, later merged with SFDCL in 2018), the West Bengal State Fishermen's Co-operative Federation Ltd. (BENFISH-the apex body of fishermen's co-operative viz.) and the FFDA. All these Corporations, apex bodies, and agencies have been engaged over the years in executing multi-dimensional activities of the Department related to a comprehensive development of fisheries and socio-economic upliftment of the fishermen community. The ICAR-Central Inland Fisheries Research Institute (ICAR-CIFRI) at Barrackpore, due to its very presence in the State, has partnered in bringing significant development in fisheries and aquaculture in the State. West Bengal has been a pioneering State on most occasions to promote evolving aquaculture technologies in all forms. During the initial period of aquaculture development in the State, ICAR-CIFRI has been instrumental in demonstrating different technologies including seed production and composite fish farming under various schemes. West Bengal had been the leading fish-producing State in the country until the emergence of aquaculture activities in the Koleru region that pushed Andhra Pradesh to the top. Among the 19 districts, South 24 Parganas is the lead fish producer followed by Paschim Medinipur, North 24 Parganas, Burdwan, and Nadia. Though the State produces 18.43 lakh tonnesof fish annually (2020-21), it imports fish from other States also to cater to the demand. At the same time, the State also sends almost 20% of its fish produced to other states.

2.3 Fisheries and aquaculture in Bihar

Bihar is a land locked state in the eastern part of the country, situated along the Ganga River. It is endowed with 93,296 ha of ponds and tanks, 2.0 lakh ha of *chaurs* and flood plain wetlands, 9000 ha of oxbow lakes or *mauns*, 26,303 ha of reservoirs, and 3200 km of rivers. Apart from the mighty Ganga River system, there are 15 rivers and rivulets drain across the State. Being an agrarian-dominant State, 88% of Bihar's 1248 lakh total population relies on agriculture, animal husbandry, and fishing for their livelihood. Inland fisheries and aquaculture have played a critical role in the livelihood of the 60.28 lakh inland fishermen population ensuring nutritional security and creating employment in the State. Unpredicted and extreme weather stresses such as the frequent occurrence of floods in the north and droughts in south Bihar have been the major challenge for aquaculture development in the State. Despite having adequate resources, the ponds and tanks available were underutilized due to poor awareness of scientific fish farming, lack of financial support, poor marketing system and lack of interest among farmers towards aquaculture However, the momentum created by the State has brought significant development in aquaculture sector during last two decade that has placed it at 5th in inland fish production and 7th place in freshwater seed production in the country.

2.4 Fisheries and aquaculture in Jharkhand

Jharkhand is another land-locked State in the eastern part of India and is enclosed by West Bengal to the east, Chhattisgarh and Uttar Pradesh to the west, Bihar to the north and Odisha to the southern part. The State has 24 districts spread over 79.716 lakh ha total geographical area, much of which lies on the Chota Nagpur Plateau.The State is endowed with enormous aquatic resources consisting of 66,348 ha tanks, 1,21,117 ha reservoirs, 4570 ha check dams and Aahars and 9,880 ha coal pits. Fisheries is an important economic activity in the state for additional employment and income generation. Of the 386.0 lakh total population, about 65-70% of people in the state consume fish. Jharkhand is a moderate fish consumption state having a per capita consumption of 10.60 kg/yr. Fisheries and aquaculture form an important livelihood option for the 1.41 lakh fishermen population living in the State. Despite being a land-locked state, it has stood at 10th position in total fish production among the States in India in 2020-21. During the last five years, Jharkhand has received significant support from the State Government besides the central aid. The Department of Fisheries has been instrumental in implementing several development plans on aquaculture in ponds, cage culture as well as improving the fish seed availability. The annual fish production of 1.9 lakh tonnes in 2017-18 has been increased to 2.57 lakh tonnes in 2021-22 with an annual growth rate of 10.02% (DoF, 2022),

3. Marine Fisheries Development

The eastern Indian maritime states viz., West Bengal and Odisha contributed 9% of India's marine fish landings in 2022. Of the two states, West Bengal (WB) dominated with 1.90 lakh tonnes forming 58.8% of the marine landings of the region, the rest 41.2% contributed by Odisha (OD). The commercially important marine resource profile is generally similar in both states with pelagic resources dominating in 2022 (48% in WB and 51% in OD), followed by demersal resources (31% in WB and 30% in OD), crustacean resources (17% in WB and 15% in OD) and molluscan resources (4% in both WB and

OD). In WB, the top five landed resources in 2022 were anchovies (18.75% of total marine landings of the state), penaeid prawns (18.58%), croakers (14.51%), pomfrets (14.26%) and Bombay duck (13.19%). In Odisha, the top landed resources were penaeid prawns (16.71%), croakers (13.06%), anchovies (9.99%), ribbonfishes (8.53%) and catfishes (7.00%). Of the three fishing sectors, the mechanized sector dominated in both States (86% in WB and 67% in OD), followed by the motorized sector (12% in WB and 29% in OD) and the artisanal/non-motorized sector (2% in WB and 4% in OD). The most productive fishing season in both states was the third quarter of the year during July-September when 43 and 32% of annual marine landings were brought onshore in WB and OD respectively. The least productive fishing season for both states was the second quarter (April-June) which coincided with the fishing ban along the east coast of India.

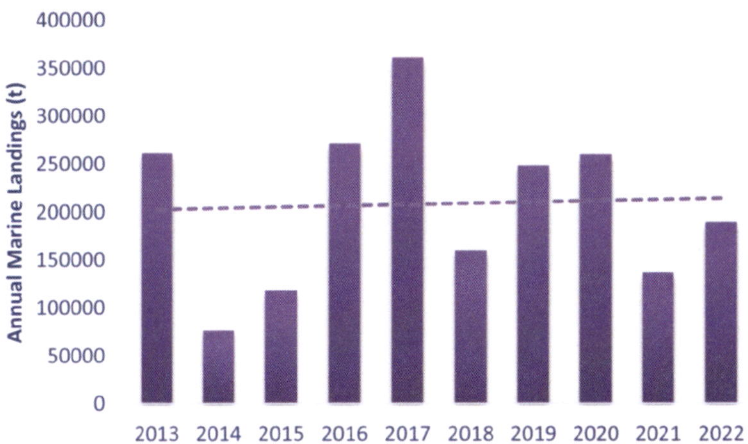

Fig. 2: Annual marine landings (t) of West Bengal from 2013-2022

West Bengal possesses 158 km long coast line. The annual average marine fish landings of the state during 2013-22 was 2.085 lakh tonnes; the annual landings ranged from a high of 3.61 lakh tonnes in 2017 to a low of 0.76 lakh tonnes in 2014 (Fig. 2). During this time period, West Bengal exhibited a steady trend in annual marine landings. The marine landings of the State were valued at 3153 crores in 2022 (Landing centre level) which formed 0.21% of the State's total GDP at current prices. West Bengal has 11054 crafts in the marine fisheries sector of which mechanized crafts were 4014 (36.3%), motorized were 6564 (59.4%) and non-motorized were 476 (4.3%). The major mechanized fishing crafts in the state were trawlers (49.9%), gillnetters (43.9%), and bagnetters (4.8%). The most dominant craft and gear in terms of marine landings in the state was the mechanized trawlers, contributing 68%

of landings, followed by mechanized gillnetters (16%) and inboard bagnetters (8%). Among the pelagic resources, the major ones landed in the state in 2022 were Bombay duck (13%), other clupeids (12%), *Coiliadussumieri* (10%), lesser sardines (9%), ribbonfish (8%), hilsa shad (7%), *Setipinna* sp. (7%), Indian mackerel (7%), *Thryssa* sp. (6%) and horse mackerel (5%). Among the demersal resources the major landings were contributed by croakers (23%), catfish (16%), other perches (10%), black pomfret (8%), soles (8%), silver pomfret (7%), Chinese pomfret (7%), threadfins (5%), threadfin bream (3%) and sharks (3%). Cuttlefish contributed the highest landings (91%) to the total molluscan landings followed by squid (9%). Within the crustacean resources highest landings was contributed by penaeid shrimp (65%) followed by crab (22%) and non-penaeid shrimp (13%).

Odisha has a coastline spanning over 480 km long coast line with 24,000 km^2 within the continental shelf. The marine fish production had declined in the State from 1,30,767 tonnes in 2007-08 to 1,20,000 tonnes in 2013-14. However, policy-driven management, infrastructure development, and improved mechanisation of the fishing fleets were the forces behind the significant development in the fish production in this sector in the later years that has led to a significant rise in marine production upto1.73 lakh tonnes in 2020-21. The annual average marine fish landings of Odisha during 2013-22 was 1.3 lakh tonnes; the annual landings ranged from a high of 1.8 lakh tonnes in 2021 to a low of 0.89 lakh tonnes in 2018 (Fig. 3). The maximum sustainable fish yield (MSY) from marine source has been set at 2.93 lakh tonnes up to a depth of 200 m off the coast. The NFDB has actively supported the development of the marine infrastructure including the establishment/modernisation of the fishing harbour and landing centres.

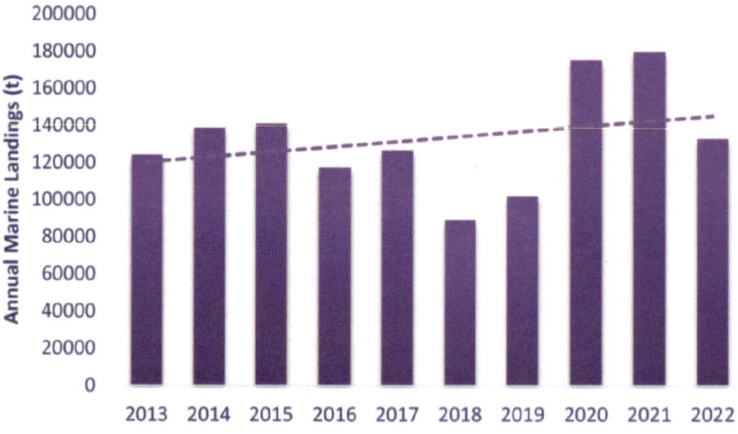

Fig. 2: Annual marine landings (t) of Odisha from 2013-2022

The marine landings of the state were valued at Rs. 2521 crores in 2022 (Landing centre level) which formed 0.39% of the state's total GDP at current prices. Odisha has 8682 crafts in the marine fisheries sector of which mechanized crafts were 1748 (20.1%), motorized were 5678 (65.4%) and non-motorized were 1256 (14.5%). The two major types of mechanized fishing crafts in the state were trawlers (79.5%) and gillnetters (20.5%). The most dominant craft and gear in terms of marine landings in the state were the mechanized multiday trawlers, contributing 62% of landings, followed by outboard gillnetters (19%). Among the pelagic resources, the major ones landed in the state in 2021 were other clupeids (18%), lesser sardines (8818 t, 10%), Indian mackerel (8210 t, 10%), ribbonfish (7859 t, 9%), *Thryssa* sp. (6776 t, 8%), other carangids (4868 t, 6%), oil sardine (4527 t, 5%), other shad (4288 t, 5%), horse mackerel (4128 t, 5%) and Bombayduck (3591 t, 4%). Among the demersal resources the major landings were contributed by croakers (36.9%), y pomfrets 16.5% (9245), catfishes 11.5% (6444 t), other perches 9.1% (5122 t), silverbellies 4.7% (2622 t), soles 4.4% (2474 t), rays 3.3% (1873 t), sharks 3.2% (1773 t), eels 3.1% (1752 t), threadfin breams 1.9% (1071 t) and threadfins 1.5% (838 t). Within the crustacean resources, the highest landings were contributed by penaeid shrimp (86.7%) followed by crab (13%), lobsters (0.2%) and non-penaeid shrimp (0.1%). Among molluscans, the cephalopods were the major resources of which cuttlefish formed 83%, squids 15% and octopus 2%. At present, there are one major and three minor fishing harbours and 69 fish landing centres (25 with jettys) supporting the marine fisheries in the state.

Stock assessment of major finfish and shellfish stocks of the north-east region (WB & OD) as assessed by ICAR-Central Marine Fisheries Research Institute (CMFRI) indicated a healthy status in the region. A total of 16 major stocks were assessed for this region, of which 8 were finfish stocks and 8 were shellfish stocks. Of the 16 stocks assessed for the north-east region, 87.5% were found to be healthy (75% of the assessed finfish stocks were healthy and 100% of shellfish stocks assessed for the region were healthy). The finfish stocks assessed for the region were that of *Coiliadussumieri* - gold spotted grenadier anchovy; *Pampus griseus* - Bengal silver pomfret; *Otolithesruber* - tiger tooth croaker; *Trichiuruslepturus* - large head hairtail; *Nemipterus japonicus* - Japanese threadfin bream; *Sardinella longiceps* - Indian oil sardine; *Euthynnusaffinis*- little tuna and *Rhinobatoslionotus* - smooth back guitarfish. The shellfish stocks assessed for this region were that of *Charybdis feriata* crucifix crab, *Portunussanguinolentus* - three-spot swimming crab; *Parapenaeopsisstylifera*- Coromandel shrimp; *Solenoceracrassicornis* - coastal mud shrimp; *Sepia aculeata* - needle cuttlefish; *Sepia pharaonis* - Pharaoh cuttlefish; *Sepiellainermis* - spineless cuttlefish and *Uroteuthis*

(*Photololigo*) *duvaucelii* - Indian squid. Of these, only the stocks of tigertooth croaker and smoothback guitarfish were assessed as overfished, rest of the stocks were assessed as sustainable. Management recommendations to improve stock status of these two resources include strict implementation of Minimum Legal Size and Maximum Legal Size (to protect gravid females of the guitarfish), fishing closure in spawning grounds during peak spawning seasons, incentive schemes to encourage return of live guitarfish, continuous monitoring programs and regular stock assessments as well as participatory management involving primary stakeholders.

4. Inland Fisheries Development

The states of eastern India are blessed with vast fisheries resources in the form of wetlands, reservoirs, rivers and estuaries.

4.1 Riverine & estuarine fisheries

West Bengal exhibits diverse topography and an intricate drainage network of 29 river basins with major area coverage by the Ganga, followed by Subarnarekha and other small river basins. The overall length of rivers and canals was estimated at 2526 km. The rivers of West Bengal can be divided into four major groups, *viz*., the North Bengal rivers (Mahananda, Teesta, Jaldhaka, Torsa), the Ganga-Bhagirathi system (Bhagirathi, Hooghly, Jalangi, Churni, Mathabhanga), the Western tributaries (Mayurakshi, Ajoy, Damodar, Darakeswar, Kansai, Silai) and the tidal creeks of Sundarbans (Saptamukhi, Jamira, Matla, Bangaduni, Gosaba, Baratala, Ichhamati). Odisha is drained by 13 major rivers and their tributaries: Mahanadi, Brahmani, Subarnarekha, Budhabalanga, Jambhira, Rushikulya, Bahuda, Indravati, Nagabali, Banshadhara, Baitarani, Nagavali and Salandi with a combined length of 2583 km in Odisha are the major rivers. The Chhotanagpur plateau of Jharkhand observed the origin of many rivers like Damodar, Subarnarekha, etc. On the other hand, Bihar is blessed with the River Ganga and her tributaries like Sone, Ghagra, Kosi, BurhiGandak, etc.

The extensive riverine network of West Bengal harbours about 239 freshwater fish species, whereas 312 marine and estuarine fish species have been recorded from Indian Sundarbans. The Hooghly-Matlah estuary, the largest estuarine system in India, is the mainstay of inland capture fisheries production in the country. The total catch of high-value Hilsa (*Tenualosailisha*), which accounts for 15-20% of total landings from this estuary, had gradually declined over the years from the highest recorded of 77,912 tonnes in 2010-11 to 20,000 tonnes in 2020-21.

Major estuaries of Odisha included Bhitarkanika (Brahmani-Baitarani estuary), Subarnarekha estuary, Mahanadi estuary, Rushikulya estuary, etc. The Chilika Lake of Odisha covering a water area of 0.79 lakh ha forms an important brackishwater resource. Recent studies recorded 148 fish species under 53 families from River Mahanadi, indicating the rich fish germplasm of Odisha rivers, with an average CPUE of 5.78 kg/fisher/day. The estuarine fish diversity has reduced from 134 species in 1996-97 to 90 species in recent years. The average estuarine catch during 1961-64 was 646.3 tonnes which increased to 3674.7 tonnes in 1996-97, drastically declining to 110 tonnes in 2005-06. Studies in river Kathajodi showed the highest fish diversity (53 species) at Nuagarh with the dominance of marine migrants. Low fish diversity, 36 and 38 species, respectively, were recorded at Naraj and Galadhari due to the reduced river flows from the Jobra barrage. Reduced river discharge followed by sedimentation was found to be the major cause of limited Hilsa fisheries in the river.

4.2 Reservoir fisheries

As per a recent survey, the fish yield from small, medium and large reservoirs in the states of West Bengal, Bihar, Odisha and Jharkhand have increased significantly through fingerling stocking and the adoption of improved management practices. However, there exists a gap in fish production and potential (Table 2), which calls for the adoption of scientific management practices and technological interventions. Stocking of advanced carp fingerlings at the rate of 1000, 500 and 300 nos/ha/yr in small, medium and large reservoirs, respectively is recommended to harness the production potential of these reservoirs. For the same, it is estimated that annually about 190 million nos of fingerlings are required.

Table 2. Present fish production and production potential of reservoirs of eastern India

State	Large		Medium		Small	
	Present yield (Kg/ ha/yr)	Potential Yield (Kg/ha/ yr)	Present yield (Kg/ ha/yr)	Potential Yield (Kg/ha/ yr)	Present yield (Kg/ ha/yr)	Potential Yield (Kg/ha/yr)
Bihar	-	-	30-40	110	90	300
Jharkhand	52	120	98	200	210	500
Odisha	30	90	85	190	290	500
West Bengal	20	100	70	175	95	400

The major issues in achieving the potential of reservoir fisheries are insufficient availability of quality seeds of desired size for stocking, inadequate fish seed raising facility, unscientific fish stocking, fluctuating water levels, habitat degradation/siltation, lack of auto recruitment, loss of biodiversity, climate change and water stress, destructive fishing, insufficient funding/ credit facility, unorganized input-output sectors, poor fisheries infrastructure, ownership issues, absence of efficient governance, institutional arrangements, and policy support, difficulty in fishing and lack of skilled/trained manpower. Strategic action plan for reservoir fisheries development included: productivity assessment and setting strategies; establishment of nursery facilities for seed production; *in-situ* fish seed raising in Pens/captive nursery to stockable size of >100 mm; stocking of fingerlings (>100 mm) for CBF and stock enhancement; strengthening of fisheries infrastructure and support services; effective value chain for marketing and enhanced income; and capacity building and knowledge strengthening of stakeholders.

The average productivity of the reservoirs in Odisha was around 9.3 kg/ha. However, the formulation of the reservoir policy and its implementation has helped in improving fish production from the open waters. Promotional initiatives in terms of long-term leasing and promoting culture-based fisheries through the continuous stocking of advanced fingerlings based on the provisions as contained in the Reservoir Fisheries Policy have increased the reservoir productivity upto 93 kg/ha.

In recent years, the promotion of cage farming in selected reservoirs of the state has opened the scope to increase fish production from the open waters. ICAR-Central Inland Fisheries Research Institute (CIFRI) has installed 110 Galvanized iron (GI) cages and a circular cage in reservoirs of Odisha for the expansion of the cage polyculture system. The state is presently promoting the use of circular cages in selected reservoirs to intensify and increase carp and other fish production.

4.3 Wetland fisheries

Floodplain wetlands are an important resource for fish production and livelihood, besides rendering other numerous ecological goods and services. These wetlands include ox-bow lakes, sloughs, meander scroll depressions, swamps, residual channels or tectonic depressions. These are widely recognized as *beel* in West Bengal and *maun/chaur* in Bihar. Bihar has a maximum area under floodplain wetlands (2,40,000 ha) followed by West Bengal (42,500 ha). However, West Bengal has been recording higher fish production (700 kg/ ha/yr) compared to Bihar (300 kg/ha/yr) owing to good governance patterns through active fishermen cooperatives and the adoption of management practices like culture-based fisheries (CBF), macrophyte management, and staggered stocking and harvesting.

Adoption of CBF, creation of infrastructures, capacity building, and assistance in input supply are important which need to be given adequate attention to achieve the production potential of 1500 and 800 kg/ha/yr in the wetlands of West Bengal and Bihar, respectively. Some of the identified issues leading to low fish production in wetlands are: habitat degradation and decline of natural fisheries owing to overexploitation; poor adoption of scientific enhancement technologies; lack or inadequate infrastructure; unorganized institutional arrangements; poor linkages; and weak financial capacity of cooperatives. Recently, ICAR-CIFRI has demonstrated integrated wetland fisheries management strategies in five wetlands of Bihar, showing an increase in fish yield from 60-190 kg/ha/yr to 87-470 kg/ha/yr. Fish production in floodplain wetlands of West Bengal increased from 150-350 kg/ha/yr to 308-3021 kg/ha/yr due to the adoption of CBF. The Fisheries Co-operative Societies (FCSs) have initiated macrophyte management, especially floating weed *Eichhornia* sp. for habitat management. Some wetlands have also laid guidelines for minimum size at capture and observe closed fishing during the breeding season to facilitate auto-recruitment.

5. Aquaculture Development

Before the beginning of the systematic fish farming activity in the 1st half of the 20th Century in the country, the vast riverine stretches were serving as the potential resources for fish supply in the eastern region. The riverine environment was also the source of fish seed supply to the aquaculture activity which was low-key during this period. Major thrust wasgiven during the last century to study the riverine ecosystem and its dynamics, fish population dynamics and the demographic environment that was dependent on such riverine environment. However, much of the efforts remained disorganized until the establishment of the Central Inland Fisheries Research Station at Barrackpore in 1947 which was later elevated to a premier research Institute the 'Central Inland Fisheries Research Institute (CIFRI)'. The establishment of CIFRI brought a systematic approach tothe development of fisheries in inland waters. Studies were initially focussed on commercially important inland fishes, and their biology including age and growth, food and feeding, reproduction, migration, etc. Further, ecosystem productivity, biodiversity, estimation of fishing efforts and landings, and fisheries-related environmental studies were other areas of focus in the later half of the last century (Silas, 2003).

Development of the aquaculture activity was also an offshoot activity from the Central Inland Fisheries Research Station that became very prominent in the later half of the Century. The eastern Indian states have a huge number of homestead ponds which formed the basis of fish culture activity till the middle of the last century. To provide adequate impetus to pond aquaculture, CIFRI established the Pond Culture Division at Cuttack, Orissa in 1949, which subsequently strengthened the base for research and development of pond aquaculture. The Division later emerged into a full-fledged institute as the Central Institute of Freshwater Aquaculture (CIFA) at Kausalyaganga, Bhubaneswar. CIFRI also initiated scientific prospecting of the carp seed collection sites in the rivers, their handling and transport for culture in ponds and reservoir stocking. While CIFRI and CIFA have been instrumental in the development of inland fisheries and aquaculture activities, the region with two coastal states also has benefited from the contribution of the Central Marine Fisheries Research Institute (CMFRI) and Central Institute of Brackishwater Aquaculture (CIBA) in the marine and brackishwater fisheries and aquaculture activities.

5.1 Freshwater Aquaculture

Post-independence era witnessed two-pronged development in the aquaculture sector, one for improving the seed quality and the other for improving the pond production in freshwater aquaculture. Stocking of seed collected from the riverine sources had its inherent problem of mixed seed and poor growth often leading to low production in the pond, Efforts were made to improve the natural habitat through the bundh breeding technique while stress was also given towards the development of hormonal induction technique for controlled breeding of the commercially important species. On the other front, the effort to improve pond productivity has led to the introduction of exotic species and development of the composite fish farming technologies.

5.1.1 Aquaculture Development in Odisha

Freshwater aquaculture in Odisha has witnessed significant improvement with an increase in the culture area and fish production. The development was promoted through district-level Fish Farmers Development Agencies (FFDAs). The inland freshwater fish production has increased significantly from 2.24 lakh tonnes in 2010 to 6.90 lakh tonnes in 2022-23. Such growth is realised through holistic development through the promotion of fish seed production and availability, financial support for seed and grow-out fish production, creation of new water bodies, diversification of the cultured species and culture systems and the overall development of the fisherman over the years. The Department of Fisheries has 78 fish seed rearing farms, 28 Indian major carp seed hatcheries and one tilapia hatchery, and the private sector has 96 hatcheries for Indian major carps and 4 hatcheries for tilapia. The fry production in the state has reached 141.31 crore in 2020-21. During the last decade, freshwater fish farming in the State has been adopted on an industrial scale in some pockets, particularly in the Puri, Ganjam and Balasore districts which has led to a considerable increase in fish production. The creation of new ponds and tanks through subsidized schemes of the Government has promoted the spread of fish farming activities to nooks and corners of the State. While the three species of Indian major carps form the main cultured species, the other major species include common carps, striped catfishes, pacu, minor carps, barbs and freshwater prawn.

5.1.2 Aquaculture Development in West Bengal

West Bengal is the only state in India, where fish have been cultivated in every kind of water bodies' including sewage water. The state has huge ponds and tank resources, mainly consisting of the homestead ponds. The composite fish farming of carps introduced in 1971 has been the mainstay in inland aquaculture practice in the State. However, over time, several alterations in the management protocol of farming have been made by the research organisations and the farmers are solely guided by their experiences and acquired expertise through the ages which are proven economical. Sewage-fed fish farming forms one of the important aquaculture activities and the State has pioneered the culture method to utilise the vast sewage-fed resources. In recent years, however, the traditional sewage-fed farming technology has substantially been modified incorporating a primary treatment process prior to use in fish ponds. Polyculture of Indian major carp along with silver carp and common carp with treated sewage water has been yielding production levels of 3.0-4.0 tonnes/ha/yr in ponds.

Fish seed production in West Bengal had started with riverine hatchling harvesting and nursing at either end of the Jessore road which links West Bengal and Bangladesh. Later, it has developed as private-sector enterprise clusters. During a later period, bundh breeding started shifting the riverine dependency of seed production towards controlled breeding. Development of the induced breeding technique in 1957 was readily accepted in the State. Further, the evolution of various hatchery models, from plastic bean to eco-hatchery models, has intensified the fish seed production activity in the State. The AICRP of the Fish Seed Prospecting operated by CIFRI during the 1960s had spearheaded the revolution in fish seed production in West Bengal. Since then, it has maintained the leading position in seed production of almost all varieties of cultured species and has been supplying seed to all parts of the country. The State also has one of the best established and developed fish seed networks in Asia. With a network of about 1500 hatcheries, West Bengal has been the largest seed-producing state in the country. Naihati and Bankura are two widely recognised fish seed markets catering to the need of the whole country.

5.1.3 Aquaculture Development in Bihar

The freshwater aquaculture resources in terms of ponds and tanks are not homogeneous in Bihar. While the maximum number of government ponds and tanks is found in Darbhanga and Madhubani districts, the highest area of private ponds is found in East Champaran (6569.39 ha) and Darbhanga (6758 ha)The traditional fish farming method was in vogue in the State due to the low awareness levels and smaller size of the majority of the ponds (<1.0 ha). Department of Fisheries, Government of Bihar has implemented several schemes to upgrade the skill of fish farmers and improve their capability for scientific fish farming.

The State has an ample number of fish seed hatcheries for seed supply to the sector. However, quality seed production protocol was lacking in the majority of these hatcheries resulting in poor quality/inbred/hybrid seed production often resulting in poor yield. Bihar is one of the first few States that successfully introduced regulatory measures on these hatcheries for quality seed production. At present, 18 carp hatcheries are operating in the State consisting of one in the government sector, three in the corporate sector, and 14 in the private sector. Promotion of the seed rearing in the open water through cage and pen culture technology has further boosted the seed production capacity. In 2020-21, with 1706.9 million fry production, the State ranks as the 7th largest seed-producing State in the country.

The major cultured species in Bihar have been the three Indian major carps alone or along with the exotic silver carp, grass carp and common carp, cultured under polyculture. Out of the 7.62 lakh tonnes produced in 2021-22, the group carp constituted 5.06 lakh tonnes while other major groups included catfish, murrels, etc. Catfish *Pangasianodonhypophthalmus* has been increasingly adopted in recent years in ponds. There has been a significant rise in inland fish production in the state which has increased from 2.8 lakh tonnes in 2010-11 to 7.62 lakh tonnes in 2021-22. Although the capture fisheries from the open waters, *chaurs* and *mauns* form a significant share in this, most of the increase was due to the rising aquaculture activity in the State. *Chaurs* and *mauns* are important resources in the State for short-term fish farming. These resources, once dependent on natural stocking, are now actively being used for seed rearing and short-term grow-out fish culture, thereby contributing a considerable share to aquaculture production. Similarly, cage farming has been introduced with a lot of impetus given by the Fisheries Department for alternative fish production from open waters.

Concerning the logistic supply, the aquaculture sector has been relying on other states for feed, aqua-medicine, and other chemicals, etc. Although there are six fish mills of different capacities operating in the State, it is sufficient to fulfill the demand. Besides, the horizontal expansion of culture areas and the shift from traditional farming to scientific practice have created more demand for aquaculture inputs in recent years.

5.1.4 Aquaculture Development in Jharkhand

The major share of the fish produced in Jharkhand came from the pond/tank and cage aquaculture. Indian major carps form the major group contributing 1.96 lakh tonnes followed by 0.21 lakh tonnes of exotic carps, mostly produced from the pond resources while*Pangasianodonhypophthalmus* has been adopted as a major species in cage farming in the open waters. At present, Jharkhand is a leading state in cage farming with the establishment and operation of around 3500 cages in operation in different reservoirs.

Since 2017-18, the State Fishery Department has implemented several developmental schemes for promoting fish seed production in the State. As a result, Jharkhand has emerged as the second largest fish seed producer in the country increasing the fish fry production from 4,150 million in 2017-18 to 11,495 million in 2020-21. The above quantity of seed is being produced from the 176 hatcheries. However, the production of quality seed has been a major concern in the State, and the Fisheries Department has been engaged in reducing the problem by improving the capacity building of the seed producer.

5.2 Aquaculture Technology Development and Adoption

The details of the development in seed production and grow-out farming in freshwater aquaculture are mentioned in the following pages.

5.2.1 Fish breeding and seed production technologies

Development of bundh breeding technique

The collection of seeds from the open-water system was the only source of seed supply to the freshwater sector until the 1950s in the country, hindering aquaculture development. The preliminary success came in the form of the development of the bundh breeding technique in the 1960s, which become popular, particularly in West Bengal (Midnapore and Bankura) and Madhya Pradesh (Nowgong in Chhattapur District). Two types of bund breeding techniqueswere followed. While the perennial type (wet bundh) was developed in West Bengal, the seasonal or dry bundh technique was developed in Madhya

Pradesh (Basavaraja, 2007). A typical Midnapore-type wet bundh was located on a gradual slope with a vast catchment area in the vicinity. It had a central deeper portion that retained water perennially where adequate stock of brood fish was maintained. During heavy rains, rainwater was entering into bundh through the inlet in the upper end submerginga major portion of the bundh and the excess water was drained through a guarded outlet (bamboo fencing) in the opposite lower end. The submerged shallow areas of the bundh (moans) wasserving as breeding grounds for stocked brooders. A dry bundh (Madhya Pradesh type) had a catchment area in the gradual slope and a depression/pond enclosed by an embankment on three sides to receive the rain water. The excess rainwater was drained through a guarded outlet on the lower side. Matured broodfish from a perennial pond were released into dry bundh preferably on cool rainy days, leading to successful spawning following heavy rain that was inundated by the shallow catchment area.

Success in the Bundh breeding technique though ensured the breeding and seed production in somewhat controlled conditions, but it still had uncertainty due to the dependence on the rainfall and many natural factors.

Development of induced breeding technique

Bundh breeding was a popular technique for fish seed production fromthe 1960s to 1980s. The quest for breeding the fish in control conditions led to greater focus on the basic studies on reproductive physiology during the first half of the 20th Century. Several types of natural and synthetic inducing agents from fish and other organisms were tried to induce the carps to breed, besides attempting to simulate the natural riverine condition for spawning induction. Some on the list include human chorionic gonadotropin (HCG), pregnant mare serum (PMS), luteinising hormone-releasing hormone (LHRH), LHRH-A, homologous and heterologous pituitary gland extracts of fishes, etc. The first successful induced breeding was achieved in 1955 in *Esomusdanricus* using catla pituitary gland extract. However, the epoch-making success in induced breeding was achieved in *Cirrhinusreba*on 10th July 1957 using carp pituitary extract at Angul farm in the Dhenkanal district of Odisha by Drs Hiralal Chudhury and K. H. Alikuni from the Pond Culture Sub-station of CIFRI at Cuttack. Subsequently, the Indian major carps (IMC) were successfully induced bred within fifteen days both at Angul and Cuttack farms of Odisha. During the subsequent period, this technique helped in the breeding of several freshwater fishes. Grass carp was bred through hypophysiation (Pituitary gland donors: IMC and exotic carps) in 1962 and followed by silver carp. Seeds were produced on a large scale and supplied to different parts of the country for culture.

Development of the induced breeding technique has also facilitated the breeding of carps in the bundh system, particularly in Dry bundh. In this method, matured brooders from perennial ponds were released in the pond of the dry bundh system. Fishes were left undisturbed for 2-3 days for acclimatization following which only 10-20% of brooders were injected with intramuscular injection of pituitary extract and an artificial water current was created by drawing water from a stored tank. After the removal of spent brooders in next morning, the eggs used to be collected using net happa and further incubated in double-walled hatching hapa (Outer happa and inner happa) for spawn collection. This traditional practice of bundh breeding was contributing a significant share of about 63% in 1980s in the carp seed supply to the sector (Basavaraja, 2007).

Development of inducing agents

The technique of induced breeding with the use of the pituitary gland in carps opened a new and better avenue for seed production. However, the effectiveness of the PG extract was largely dependent on the freshness, potency and preservation techniques of the gland. The use of this extractwas subjected to problems of inconsistency in the potency, and further, the spawning was largely dependent on favourable environmental conditions. It involved two-time injections and too much handling of the brooder was leading to post-spawning brooder mortality. Since it involved the collection of large amounts of pituitary glands from matured fish and also had problems of storage, often there was a shortage in the availability of the gland. Some attempts were made to preserve the gland through ampouling, but had many operational constraints. To overcome such a problem, the research activity during 1970s was oriented towards the development of synthetic hormones. Several attempts have been made in the country to develop a synthetic version of the carp gonadotropin-releasing hormone. Successful attempts were made to use dopamine antagonists (domperidon, pimozide) along with the PG extracts to increase the indigenous surge of the gonadotropin along with the exogenous hormone sources to induce spawning. During 1980s, the Syndel Laboratory (India) Pvt. Ltd., Mumbai brought the synthetic inducing agent 'Ovaprim', which was a synthetic analogue of the Salmon gonadotropin-releasing hormone mixed with domperidone. The requirement of a single dose of Ovaprim reduced the broodfish handling, increased the post-breeding brooder recovery and most importantly improved all the breeding parameters such as spawning fecundity, rate of fertilisation, % hatching and spawn recovery. Following the success of the Ovaprim, several synthetic inducing gents followed suit and revolutionised the carp hatchery operation and large-scale seed production. The list includes Ovatide (SGnRH-A+pimozide and later SGnRH-A+domperidone), Wova- FH (SGnRH-A+ domperidone), Gonopro, etc.

The development of synthetic inducing agents has also helped in developing induced breeding protocols for many other economically important groups of species such as minor carps, catfishes, murrels and ornamental fishes. The use of LHRH is also in vogue for breeding catfishes. Hormonal implant techniques have also been successfully developed for the induced breeding of catfishes and murrel. Today, the induced breeding technique has been standardised for more than 40 commercially important freshwater food fishes apart from many ornamental fishes.

The general practice of carp breeding was to breed the matured individuals once during the season and the hatchery owners had to maintain a large population of the broodstock to cater to the seed demand which was taxing extra expenditure for the broodfish maintenance. Efforts were made to reduce such problemsby exploring the possibility of using the same brooders multiple times during the same season. During the later part of 1990s, ICAR-CIFA succeeded in developing a protocol for multiple spawning, i.e. breeding the same brooders as many as four times during the same season (March, May, July and September) through effective broodstock care. With this method, average spawn production increased from 0.05-0.06 million/kg in March to 0.15-0.20 million/ kg female body weight from the 1st to 3rd (monsoon) breeding but reduced in the 4th one. However, the multiple spawning techniques demonstrated 2-3 folds higher spawn recovery over conventional single breeding. Standardisation of technique for cryopreservation of the carp milt during 1990s was another significant development that has reduced the requirement of male broodstock population in the hatchery. Cryopreserved carp milt has been used as an effective tool for stock upgradation, besides overcoming the problem of inbreeding deficiency in many hatcheries.

Environment manipulation for off-season breeding of carp was another line of development that took place during the last decade (2010s) with partial success. However, the development of the broodstock diet CIFABROOD™ in 2010s was a major milestone that ensured early broodstock maturation. The availability of this broodstock diet on a commercial scale at present is facilitating early breeding in the carps. Many farmers in recent years are also able to achieve breeding in the carps well before the monsoon months with the combined use of groundwater (lower temperature) in the broodstock pond as well as in breeding tanks to ensure early availability of seed.

Development of hatchery models

With the progress of research in fish breeding, the various models of hatchery systems had also evolved from the simple hatching pit to the modern eco-hatchery that operates at commercial scale. However, the focus till 1990s was mainly on the seed production of the major carps, barring the sporadic attempt to develop the hatchery model for the magur, the then next important species. Of the different hatchery models, the Eco-hatchery model is continued nowadays for mass-scale carp seed production whereas the jar hatchery system (glass, concrete types) is also in use in some hatcheries and operated for the low scale of production. More than 3000 eco-hatcheries are operating in the country. Of late the FRP hatchery system for carp was developed by CIFA for small-scale decentralized seed production. Being operated on the principle of eco-hatchery, it has the added advantage of transportation which enables its reach to remote areas for carrying out the breeding operation. More than 300 FRP have been set up in various parts of the country to carry out carp breeding. The same hatchery system has also proven to be effective for the breeding of several minor carp species. Some hatcheries in the country use the updated version of the eco-hatchery model. While the eco-hatchery makes use of separate varied size tanks for breeding and egg incubation, the new model includes a two-in-one version, i.e. the egg incubation tank is made large enough to act as the breeding tank. In such a system, the spent brood fish are removed after spawning and incubation of the releasedegg is continued till the development to spawn stage. The use of hapa for carp breeding is also being practiced by a few farmers nowadays for seed production on a limited scale.

With the emphasis on species diversification, there was a need to develop seed production technology for the new species. The hatchery model for magur was developed during 1990s and mainly consisted of a small circular breeding tank and a flow-through system for the incubation of the egg. Later in 2010s, the FRP model of magur hatchery was developed at ICAR-CIFA and had been supplied to farmers over the years.

The development of the seed production technology for freshwater prawn was another milestone in the freshwater sector. Following the successful closure of the life cycle of the giant freshwater prawn (*Macrobrachiumrosenbergii*) in captivity, a hatchery model was developed by ICAR-CIFA. The model basically involved the maturation of the brooders in small ponds followed by spawning in concrete or FRP tanks. The larval-rearing activity is carried out in tank system (concrete or FRP type). Following the success of this hatchery model, the hatchery system was developed for the two other prawn species *M. malcolmsonii* (Indian River prawn) and *M. gangeticum* (Ganga river prawn) also at ICAR-CIFA.

The striped catfish (*Pangasianodonhypophthalmus*) gained its importance as an important cultivable species next to the carp group. The seed demand during the initial phase of its entry into our culture system was mostly catered from the neighbouring country through illegal trade. However, increased seed demand led to the development of an indigenous hatchery model for this species, mostly set up in West Bengal. But nowadays, hatchery for this species is found in many states of the country including, Andhra Pradesh, Chhattisgarh, Jharkhand, Bihar and Odisha. Pabda (*Ompok*spp) is another small catfish that has been adopted as an important cultured species particularly in West Bengal and a few Northeastern states due to its high consumer preference and premium market price. Specialised FRP hatchery models have been developed in recent years for the breeding of this species.

Breeding and seed production technologies for a few other important cultivable species have also been developed in recent years which include the murrels, climbing perch, singhi, yellow catfish, Mahandirita, tengra, etc. However, simple concrete or FRP tank system is used for most of these fishes to facilitate egg spawning. Since murrels and climbing perch lay floating eggs, tank system is also used for the incubation of the fertilised egg. But due to the sticky eggs being produced by the above catfishes, flow-through systems using small plastic tubs, similar to the magur hatchery system, are also effectively used for their egg incubation.

Development in the fish seed rearing technology

Development of the seed rearing technique was the other concurrent development that took place along with the induced breeding technology to cater to the ever-growing demand for seed for the sector. With the development of induced breeding, riverine seed collection was gradually pushed to oblivion. Efforts were made in the long term to refine the seed rearing technology concerning stocking density, species ratio, and use of varied types of inputs such as organic manure, inorganic fertilisers, lime, feed ingredients, etc. to increase their efficiency in terms of seed production. Effective management of the pond ecosystem was given prime importance during the development process. Most of the basic technologies of carp seed production are credited to the efforts of the Pond Culture Division of CIFRI at Cuttack during 1970s to 1980s although other research institutions in the country have contributedto the endeavour. Seed rearing has been an evolving technology during the last 70 years with more refinement brought in the use of different kinds of inputs, species combination as well as density and species ratio. Until the beginning of this century, earthen pond was the basic resource used for seed rearing of carps and it was often marred with poor seed survival and growth. However, in recent years, tank systems are effectively used for seed rearing not only for carps, but also for many new species. The better-controlled environment in the tank system has made it possible to rear seeds at higher density with the realisation of higher seed survival, growth and efficient use of inputs.

Seed rearing in carps is basically divided into two phases, i.e. (a) nursery phase where the spawn of 6-7 mm size are grown at 5-10 million/ha (500-1000 spawn/m^2) to fry of 20-25 mm within a period of 15-20 days, and (b) rearing phase where the fry are reared at 0.2-0.3 million/ha (20-30 fry/m^2) to 6-10 cm size called fingerling in 75-90 days. The basic rearing technology involves pre-stocking pond preparation (weed clearance, pH correction, manuring/fertilisation, eradication of unwanted fishes and aquatic insects); seed stocking (maintenance of appropriate stocking density and species ratio, acclimatisation); post-stocking pond management (maintenance of natural productivity through manuring/fertilisation, supplementary feeding, health management); harvesting and seed transport (conditioning, packaging and transport condition). Innumerable studies have been made in refining every step of seed-rearing activities over the last 70 years. Further studies on the water exchange and aeration requirement, feeding rate and frequency have been studied. Packages of practices have been developed for the nursery as well as the rearing phase of the carp from time to time which are still evolving. With popularisation of the intensive cropping pattern such as multiple stocking-multiple harvest and multiple cropping, there has been an increasing demand for larger seed for grow-out stocking. This has prompted the developmentof another tier of seed growers who raise stunted juveniles for a round-the-year supply of larger seed for grow-out operations.

Seed rearing techniques for species other than carps are mostly species-specific and are evolving over the years with regard to stocking density and input requirement. For example, most of the catfish during early life stages show cannibalism and hence require periodic size segregation to improve seed survival. A similar technique is also being used for the rearing of the murrel seed. The early larval stage of magur requires lower water depth (8-10 cm) in the tanks as the larva has to travel the water column to the surface for obligatory airbreathing. Indoor rearing is done during the 1st phase of rearing @ 2000-3000 nos/m^2 for12-14 days in a well-aerated tank to reach 10-12 mm fry (30-40 mg) size. The 2nd phase of rearing for 30 days is carried out at 200-300 no/m^2 densityin outdoor tanks provided with a 6-8 cm soil base to reach the fingerling size of 3-4 cm (0.8-1.0 g). In the case of freshwater prawn, brackishwater (11-13 ‰ salinity) is required to complete the 22-26 days of larval rearing in two phases, i.e., initial phase of 10 days (zoea I to V or VI) at 200-300 larvae/l density and the next phase of 12-16 days at @ 50-60 larvae/l density to complete the 11th zoea stage to reach the 1st post larval stage. Unlike the carp seed, live food (*Artemia* cyst) is a prime requirement for the successful larval rearing of prawn. Further, post-larvae are acclimatized to freshwater at a high stocking density of 2000-5000/m^2 for 10-15 days.

5.2.2 Development of grow-out fish culture technique

Development of the aquaculture sector took momentum during the post-independent period with a lot of impetus given from the government as well as research Institutes for the basic studies aimed at artificial propagation of the fishes, the introduction of exotic fast-growing species as well as popularization of the fish production technology through widespread demonstration of fish farming. By then, Fisheries Departments had been created in Bengal, Bihar, Odisha and Madhya Pradesh. These Departments have played a significant role in the research and development of the aquaculture sector. Similarly, the CIFRI spearheaded the basic and applied research for the development of the aquaculture sector in the region as well as the country during the post-independent period.

Introduction of exotic species

The introduction of new species with a higher growth rate was one of the initial strategies followed in the post-independence era for increasing fish production in ponds and natural waters. The 1950s witnessed the introduction of several exotic fishes into our culture system. In 1952, *Tilapia mossambica* was brought from Bangkok and introduced in Mandapam. The scale carp (*Cyprinus carpio communis*) from Bangkok was introduced into the pond culture system at Cuttack in 1957. In the same year, crucian carp golden carp (*Carassius carassius*) was introduced from central Europe in the Ooty Lake and later transplanted to the Nilgiris and Sunkesula fish farm of Andhra Pradesh. The exotic grass carp (*Ctenopharyngodonidella*) and silver carp (*Hypopthalmichthys molitrix*) were brought from Hong Kong and introduced at the Pond Culture Division of CIFRI, Cuttack (Odisha) in 1959. The introduction of the three exotic species silver carp, grass carp and common carp helped in increasing the productivity of the pond significantly due to their higher growth rate with the advantage of the higher temperature regime. Later, this exotic group formed an important component of the carp polyculture system, and till date, contributing significantly to freshwater fish production in the country.

Development and demonstration of composite fish culture

Development of the composite fish farming technology led CIFRI to launch the National Demonstration Project (NDP) in 1965 for technology demonstration in farmers' fields. The foundation for the blue revolution for the country was led by CIFRI launching the All-India Coordinated Research Projects on "Composite Fish Culture", "Seed Prospecting and Seed Production of carps" 'Air-breathing Fish Culture' 'Brackishwater Fish Culture' and "Ecology of

Reservoirs'. Later, the first two AICRPs were merged into one and renamed as 'Composite Fish Culture and Fish Seed Production' which was operated initially at six centres, *i.e.* in Kalyani (West Bengal), Sunkeshula/Badampuri (Andhra Pradesh), Bhavani Sagar (Tamil Nadu), Hadapsar (Maharastra), Jaunpur (Uttar Pradesh) and Karnal (Haryana), but later extended to other states also. Through adaptive research, more than 3 t/ha/6-months could be achieved in the 1st year which subsequently raised to 8 t/ha/10-months to 10 t/ha/yr at Haryana and Maharashtra Centers, respectively against the projected target of 3 t/ha/yr in five years. The CIFRI also had demonstrated the composite fish farming technology and integrated farming on a large scale through the extended version of NDP: the Operational Research Project (ORP, 1974-75). Later, the CIFRI-IDRC Project on Rural Aquaculture undertook ensured massive technology transfer in West Bengal and Odisha demonstrating 4-6 t/ha/year production level in private ponds. The composite fish culture technique has been refined to make it more productive and farmers-friendly during the transitional phase of aquaculture development. At present, production levels of 3-5 t/ha and 8-10 t/ha/year of carps under modified extensive and semi-intensive polyculture systems, respectively have become common in the farmers' fields.

Species diversification

During the course of aquaculture development, species diversification is stressed upon to increase productivity and farm income. Several new species, with or without the government's approval for culture, have been brought into the culture system. African catfish (*Clarias gariepinus*) had been one of the first illegally brought species to the country during 1990s. Due to its exceptional voracious feeding habit, resistance to extreme conditions and higher growth rate, it thrived well in the Indian waters and achieved a widespread culture practice within a short period. But later, the Government of India banned the culture of this species because of its highly carnivorous food habit that leads to ecological consequencesfor the indigenous species. However, during the short span of 20 years, this species has entered the natural waters and captured the niche of many derelict waters and posing a threat to the indigenous fish fauna.

Unlike the African catfish, the introduction of striped catfish, *Pangasianodonhypophthalmus,* has proven beneficial in the aquaculture system. Though the species also had an illegal introduction during 1990s, its higher growth rate and feasibility of high-density grow-out culture increased its popularity among the farmers, leading to an increased spread in the culture system. The species got a Government nod for culture in 2005. The potential of the species to achieve an average production level of 20 t/ha under monoculture and the feasibilityof farming in feed-based carp polyculture systems have been helping its large-scale adoption over these years. The production of the species has increased consistently and at present it forms the second leading cultured species after the Indian major carp group. It is also considered a prime choice for cage culture in open waters.

To support the farmers involved in the farming of common carpas a component species in carp polyculture systems, in 2000, an improved Hungarian strain of common carp, popularly called Amur carp, was brought to Karnataka under DFID carp genetics project - Genetic Improvement Programme of Common Carp in Karnataka', and after rigorous evaluation, it was released for commercial production. The seed of the species produced from this centre has been propagated to many multiplier hatcheries for further transmission to the farmers. The new strain with a higher growth rate, slimmer body and lower fat deposition has increased popularity among fish farmers across the country.

Tilapia (*Oreochromis niloticus*) is the other species that got official approval for culture in the country in 2009, but with certain riders such as the culture of only mono sex (male) and sufficient measures for prevention of its escape to the natural waters. The culture of Nile tilapia or its improved strains/hybrids are officially permitted.The government of India in 2009 permitted RGCA, the research arm of MPEDA, and three other Farmers' groups for the seed production and farming of tilapia (GIFT/golden tilapia) following the guidelines developed by the National Committee on Introduction of Exotic Aquatic Species into Indian waters. MPEDA was instrumental in introducing GIFT strain in the culture system to boost its exports. Large-scale seed of GIFT produced at RGCA are being supplied to different states. At present, it also forms a promising species in the biofloc system.

The silver barb *Puntius gonionotus* was introduced in the country in 1972 from Indonesia, but somehow could not draw the attention of the researchers till 1990s. But later, it has gained importance and is being considered an important component due to its auto-breeding habit, high flesh-to-bone ratio, and increasing consumer preference.

The red-bellied pacu, *Piaractusbrachypomus*had entered into the culture system of the country in 2003-2004 through the ornamental trade. Though the species is yet to get Government approval for culture as food fish, it has already achieved considerable spread in the culture system and contributes a sizeable amount to the culture production basket of several States.

Grow-outproduction technologies of other commercially important species

Culture of the three IMC (catla, rohu and mrigal) and three exotic major carps (silver carp, grass carp and common carp) have been the mainstay of aquaculture in India contributing a share of 70-80% of the inland production over the years. However, in recent years, contribution of the exotic carps has gone down. Striped catfish has become the next important cultured species in the culture system, specifically in Andhra Pradesh, Chhattisgarh, Jharkhand, Bihar and Uttar Pradesh. The yield level has often reached 40 t/ha with the intensive monoculture setup. Though pacu is yet to be approved as a cultivable species in the country, its culture area has increased during the last 10 years, particularly in Andhra Pradesh with present coverage of more than 2500 ha. There has been a growing interest inthe culture of improved variety of mono-sex tilapia and red tilapia which have received the nod as cultivable species with certain restrictions. While pacu is being cultured mostly as a component species in the carp polyculture system, tilapia has been widely adopted as the species of choice for the biofloc system. Monoculture of magur (*Clarias magur*) and butter catfish (*Ompokpabda, O. bimaculatus*) are gaining increasing popularity in the North Eastern states with 1.0-1.5 t/ha production range in both species. Of late, stinging catfish (*Heteropneustesfossilis*) has gained popularity as a culture species for the biofloc-based system. Murrels are popular in West Bengal and south Indian states like Andhra Pradesh, Telangana, Kerala and Tami Nadu. Recent success in the mass-scale seed production technology of *Channa striata* at ICAR-CFA has opened the scope for its expansion in the culture system.

During 2000-2005, there had been a rise in culture coverage of freshwater prawn *M. rosenbergii* and *M. malcolmsonii* in the sector, partly due to the setback in brackishwater shrimp farming. The sector had witnessed a steady increase in freshwater prawn production to 35,000 tonnes in 2005. However, the popularization of *Letopenaeusvannamei* as a highly productive and profit-making crop has brought the coverage of freshwater prawn down during the last decade. However, freshwater prawn farming has shown signs of revival again and is expected to spread in the coastal region in the coming days due to the rising problem in the culture of *L. vannamei*. The culture of freshwater mussel *Lamellidenmarginalis* is becoming popular as a cash crop for freshwater pearl production in the coastal as well as landlocked states following a phenomenal increase in the interest of farmers for its culture in recent years.

The grow-out farming practices have evolved depending on the market forces and regional preferences with changes in species composition, stocking density and input use. The bi-species culture of rohu (80-90%) and catla (10-20%) with stocking of larger sized seed (100-200 g) has evolved in the Koleru lake region of Andhra Pradesh, the major fish production hub of the country. Whereas, the composite culture of the three or six major carp species is still in vogue in many states. Following greater stress on species diversification, there has been the introduction of many new potential exotic species such as the striped catfish and red-bellied pacu which are farmed in monoculture or polyculture with carps. Modern aquaculture practice is transforming towards specialised and precision farming with a trend towards 'fish fattening process' where the stocking size ranges up to 400 g to shorten the crop duration to four to six months to reach marketable size. There has been a change in cropping patterns aimed at the effective utilisation of the production potential of the ponds and tanks. Different cropping methods such as single stock (more density)-multiple harvest, multiple stock-multiple harvest and multiple cropping have been devised and evaluated with the realisation of higher fish production than the traditional single stocking single harvest method. Inter-cropping of several minor carps has been proven to be useful for better utilization of seasonal ponds with higher returns. All these cropping patterns not only lead to better yield but also increase the investment capacity of the farmers through the scope of interim harvest, ploughing-back of investment and reducing the risk for the investment.

5.2.3 Development in the non-conventional farming systems

During the last seventy years of aquaculture development, diversified systems of fish farming have been developed apart from the conventional system of farming in ponds and tanks to use alternative resources and inputs for fish protein production. The list includes integrated farming systems, rice-fish farming systems, sewage-fed fish farming, cage culture, pen culture, etc.

Integrated fish farming system

Based on the principle of efficient nutrient recycling and yield enhancement, integrated fish farming involves farming several compatible animal and plant components with fish as the main component. The selection of the components for integration is largely based on their regional importance. Several integrated farming models have been developed over the years and are practised on a regional basis. For example, fish-cum-pig integration is quite popular in the Northeastern states while a large chunk of rice-fish farming systems is found in West Bengal, Assam and a few other northeastern states. Fish-cum-duck and fish-cum poultry are also common integration models found in different parts of the country. Integration models using cereal and tuber crops, horticulture, poultry, duckery, dairying, mushroom along with fish have been successfully demonstrated. Similarly, pig-cum-fish culture, duck-cum-fish culture and paddy-cum-fish culture have been popularised by the Kalyani Centre of ICAR-CIFA in West Bengal and NEH region. Integration of high-valued horticulture crops such as strawberries and other seasonal fruits on pond dyke with indigenous fish species in the pond has been accepted as a popular model in West Bengal.

With emphasis given to ecosystem health, integrated farming models are increasing their popularity in recent years due to their holistic approach to crop production with efficient nutrient recycling and resource utilisation. Such type of system has become particularly useful for the utilisation of the small household ponds found in rural areas. Besides providing a substantial income source, it also helps to secure the nutrition security of small and marginal farmers and generate alternate employment opportunities.

Rice fish farming in lowland rice fields

Fish farming in the lowland rice ecosystem has been a traditional activity in coastal areas as well as land-locked states of the North Eastern Region. Several Institutes under the ICAR have worked on developing suitable rice-fish models over the years. The notable is the models developed by ICAR-NRRI over these years. Components like apiculture, floriculture, mushroom, seasonal vegetables, horticulture crops and agro-forestry are some of the common components selected based on suitability along with fish in such integration. A typical model has a fish refuge at one end and two trenches on either side of the plot's length. Apart from growing the regular rice crop in the inundated lowlands during *kharif* season concurrently stocking with fish fry/fingerling, such a system incorporates seasonal vegetables and fruit crops, apiary and mushroom on dyke; and pulses/watermelon in the main field during rice fallow season. It has not only shown an increase in rice yield upto 35%, but also controls several rice pests and diseases as well as weeds, and minimises use of insecticides and weedicides. An average yield of 2.6 tonnes of rice and 600 kg of fish has been reported from a one-hectare rice-fish farm, besides the production of other food crops and vegetables which not only adds to the farm income but also creates 200-250 man-days of employment. Such models have been successfully adopted in several parts of Odisha and Assam as sustainable farming systems by the farmers.

Wastewater-fed Aquaculture

Fish production from sewage-fed impoundments has been a practice that evolved since 1930s and was largely practised by the fish farmers of West Bengal. However, stress had been given by CIFRI as early as 1975 to improve such farming systems for efficient utilisation of the resources to harvest fish protein. The Rahara Centre of CIFRI was dedicated to the study of the various kind of sewage, their classification, possible level of utilisation for fish farming and the after-effect of the sewage on fish as well as in the food chain. After the reorganization of the fisheries research Institutes, this wing of CIFRI is functioning as the Regional Research Centres (RRC) of CIFA, and carrying out dedicated research on wastewater utilization and management for fish farming. This Centre has carried out extensive studies on sewage characterisation and their seed as well as grow-out fish production potential both in raw and diluted form. Various species combinations and species ratios have been used to raise them in varied dilutions of sewage to standardise the fish farming protocols in such systems. Fish production levels of over 4.0 t/ha have been demonstrated in the monoculture of *C. reba* and *L. bata*and also polyculture systems in such sewage-fed ponds. The use of sewage in paddy-cum-fish farming systems is another area that has been proven to enhance rice production.

Cage and pen culture

Enclosure aquaculture (cage and pen culture) has been a promising intervention to increase the fish production potential of large water bodies. Over these years, ICAR-CIFRI has been instrumental in standardizing the technology for cage and pen farming which included aspects of cage design, enclosure (net) material, identification of suitable species combination and stocking density in cages, study on population dynamics and carrying capacity of the open waters, and impact of cage farming on the water ecosystem, etc. It has demonstrated stock enhancement through pen culture in selected wetlands of Assam which had later been adopted in many *chaurs* and *moan* resources of the eastern Indian States including Bihar, Chhattisgarh, Jharkhand, and flood plains of Assam and West Bengal. Several schemes have been brought by NFDB with the technology partnership of ICAR-CIFRI and implemented through State Fisheries Departments to increase cage farming activities in open waters. Such activities are implemented with an integrated approach as an alternative livelihood and income generation programme under the Blue Revolution Scheme of the Government of India. With the financial support of the RKVY scheme, Chhattisgarh was the pioneer state to start commercial cage farming in 2007-08 in the rivers and reservoirs system. Production of >3.0 tonnes *P. hypophthalmus*/cage of 6 m × 4 m × 4 m is achieved in 6-8 months of rearing. Now cage farming technology has been adopted in the reservoir systems of many states including Jharkhand, Bihar,and Odisha. Cage farming has also been given impetus in the recently implemented PMMSY scheme sponsored by the Central Government and the scheme has been extended to all the States for the development of open-water fisheries.

5.3 Brackishwater aquaculture and Mariculture Development

The brackishwater aquaculture development in the country, including that in eastern states like Odisha and West Bengal is steered by the adoption of several technologies ICAR-Central Institute of Brackishwater Aquaculture (CIBA) and other scientific organizations. ICAR-CIBA has standardized the technological innovations in coastal aquaculture disseminating the technology across the country. The institute has established hatcheries for the production of diversified fin fish species such as seabass (*Lates calcarifer*), milkfish (*Chanoschanos*), pearlspot (*Etroplussuratensis*), and low value species like *Mystusgulio*. With respect to crustacean culture, the institute has been prioritizing the culture of Indian white shrimp, *P. indicus* and has initiated a selective breeding program for this native species as a complementary species to *P. vannamei*. Nursery technology and multi-phase farming including biofloc technology developed and refined by the institute has been adopted by several farmers in coastal India. The Institute has also developed methodology for assessing suitable cultivable areas for aquaculture through GIS mapping and is currently adopted by different states.

West Bengal possesses vast brackishwater resources of about 2.1 lakh hectares. About 28% of the potential brackishwater areas in the state are under use. The state has been the leader in *Penaeus monodon* production in the country. The total inland production of West Bengal was 1.65 million tonnes in the year 2022 which is the second position in the nation. However, the majority of this production comes from inland capture fisheries and shrimp farming. There is wide scope for increasing the brackishwater aquaculture development by judicious use of the resources. In West Bengal brackishwater aquaculture is still practiced in the age-old traditional system in the bheries with a production of 500-700Kg/ha/yr. The major brackishwater area is mainly distributed in the districts, Purba Medinipur, South and North 24 Parganas. Identification of the potential area from the actual resources is the first step towards the development of the brackishwater aquaculture development in the state. In West Bengal majority of the farmland holdings are smaller and there are no proper leasing policies at present especially for vast areas like bheries. Policy decisions and devising programmes like cluster farming can help to tackle the issue. Physico-chemical parameters and events caused by frequent natural calamities have to be taken into consideration for the identification of suitable land. The brackishwater culture technology to be adopted has to be decided on the basis of these parameters and carrying capacity. In India brackishwater aquaculture is synonymous with shrimp farming, however, not all locations may be suitable for shrimp farming. The majority of the brackishwater wetlands in West Bengal are ecologically diverse and maintaining proper biosecurity may be a challenge. Shrimp culture is broadly accepted by aqua farmers of West Bengal due to its high economic return. There is vast scope for sustainable development of brackishwater aquaculture in the region to meet the livelihood demand utilizing the unused and underused areas and adopting advanced farming practices. ICAR-CIBA has developed several finfish-based polyculture technologies which may be adopted to improve the productivity of the traditional systems and improve the economic returns sustainably. The species composition in these technologies is selected in such a way that the economic returns can be maximized. Apart from this, the combination of herbivorous and omnivorous species reduces the feed demand which is also important for sustainability in the long run. Adopting diversified technologies can bring up the projected productivity of 1.5-3 tonnes/ha in 6-10 months in traditional systems.Ecosystem-based farming models are best suited for the development of brackishwater aquaculture in West Bengal taking into consideration the various challenges. Challenges faced by Sundarbans aquafarmers need to be tackled by appropriate management tools like social mobilization of aqua producers, technology assessment and refinement, participatory planning, and capacity building of key stakeholders.

Odisha has 4.3 lakh ha of brackishwater resources out of which 0.33 lakh ha are suitable for farming. There are 0.08 lakh ha of backwaters and 2.98 lakh ha of estuaries besides the Chilka Lagoon extending over 0.79 lakh ha. The brackishwater farming in the State has been almost synonymous with shrimp farming over the years with traces of finfish farming in the sector. The 1990s witnessed boom in the farming of *Penaeus monodon* with high-density farming and production level often reaching 6-8 t/ha. However, the white spot virus (WSHMV) epidemic has obstructed the pace of development and transformed the sector into very low-density farming. In 2012-13, the average yield was only 1300 kg/ha/yr and the total production was 13,227 tonnes. However, the last two decades have witnessed an increase in the farming of *Litopenaeusvanamei* in the State with brackishwater production, mostly of shrimp, reaching 1.25 lakh tonnes in 2020-21 with the culture area spreading to nearly 18,000 ha. ICAR-CIBA has successfully demonstrated a culture system for *P. indicus* with a production of 6-7 tons/ha in Balasore, Odisha during the year 2016-18. There are 26 private and one government shrimp hatcheries operating in the state with present production 1945.5 million post-larvae.

Fish production through marine cage farming has also been started in Odisha. Successful demonstration of fish farming in cages off the Balasore coast by ICAR-CMFRI with the active support of NFDB and the State Fishery Department has paved the way for expansion of this activity along the coast.

5.4 Development in fish feed production and health management

Successful efforts have been made to study the nutrition requirement of the fish to formulate balanced diets which were transferred to the feed industry for production of commercial feed. Over the years, many significant developments have taken place in the area of fish feed development. The last decade has witnessed the establishment of a large number of feed mills across the country, particularly in Andhra Pradesh, Odisha, West Bengal and Chhattisgarh. Floating, sinking and slow-sinking forms of commercial feeds in varied pellet sizes are now available in the market.

Research efforts in the health management sector have increased over the years following the emergence of several types of diseases consequent upon the intensification of the farming system. Similarly, other health problems such as poor growth and mortality due to water quality issues, often ignored earlier in culture ponds, have drawn increased attention to the development of better management practices (BMP). Research efforts in the sector have led to the development of several diagnostic kits for early warning and detection of disease and chemicals and therapeutics for disease control. The development of CIFAX, the chemical formulation by ICAR-CIFA was a landmark achievement for effective control of EUS, which had a devasting effect on the fish farming activity during 1990s. Disease surveillance in the sector is in place to trace and track emerging pathogenic problems and efforts are being made for their prevention and cure through research intervention. A National Surveillance Programme on Aquatic Animal Diseases (NSPAAD) has been operational in 20 States and 2 UTs with the active participation of 29 partner organisations. Referral Laboratories for Level-III diagnosis have been established in the country to support an 'Emergency response system' in the aquaculture sector.

6. Epilogue

The blue revolution in India,which started in the 1980s, has led to an increase in the total fish production from 2.44million tonnes in 1980 to the tune of 16.25million tonnes in 2021-22. Such development has been achieved with all-round support in terms of policy formulation and implementation; financial support and incentives; improvement in logistics and infrastructure facilities; improvement in gears, crafts, and techniques for the capture fisheries; development of aquaculture practices; promotion of alternate species and systems and the concurrent adoption of natural resource management; and improvement in fish marketing. The presence of the two major research institutes *viz.*, ICAR-CIFA and ICAR-CIFRI in the regionhas been an added advantage to these states for easy access to the scientific technologies. Significant rise in the fish production and export of fish and shellfish touching new heights have emphasized the relevance of the sectorto the country's economy. At present the fish production methods are being integrated with the modern tools of SMART and IoT-based interfaces. The fisheries sector is expected to come up with timely strategies to cope with the future challenges of increased fish demand, meeting the selective consumers' choices, production of safe and quality fish, and higher export earnings, etc. All these have to be done with the odds of land and water scarcity, competition for raw materials from other agricultural sectors, labour shortage and changing climatic regimes. There has been an increased tendency of intensification oriented towards producing more fish, which has posed environmental concerns. Therefore, the development machinery must be proactive and committed to adopting strategies and suitable action plans, and viable programmes for the development of the fisheries and aquaculture sectors.

References

Basavaraja, N. 2007. Freshwater fish seed resources in India, pp. 267–327. In: M.G. Bondad-Reantaso (Ed.). Assessment of freshwater fish seed resources for sustainable aquaculture. FAO Fisheries Technical Paper. No. 501. Rome, FAO. 2007. 628p.

CMFRI, FRAEED (2023) Marine Fish Landings in India - 2022. Technical Report. ICAR-Central Marine Fisheries Research Institute, Kochi.

CMFRI, Kochi (2023) Marine Fish Stock Status of India, 2022. CMFRI Booklet Series No.32/2023 (32). ICAR- Central Marine Fisheries Research Institute, Kochi, pp. 1-22.

DoF, 2022. Hand Book of Fisheries Statistics, Department of Fisheries, Ministry of Fisheries, Animal Husbandry & Dairying, Govt. of India.

Silas, E.G., 2003. History and development of fisheries research in India. Journal of Bombay Natural History Society 100 (2&3): 502–520.

3

Livestock - A Lifeline for Small and Marginal Farmers of Eastern India

Kumaresh Behera, Bhagirathi Panigrahi and Susen Kumar Panda

College of Veterinary Science & Animal Husbandry, Odisha University of Agriculture and Technology, Bhubaneswar

1. Introduction

India has 536.76 million livestock as per 20th Livestock Census 2019. Livestock plays a vital role in improving the socio-economic conditions of rural masses especially for landless and marginal farmers; as well as an important role in the national economy. Livestock production and agriculture are closely related and vital for the overall food security. Agriculture diversification through animal husbandry is one of the primary factors of growth in rural incomes. A higher public investment in livestock sector is the need of the hour for doubling farmers' income. Future economic growth would have to come from improvements in animal productivity. India is bestowed with a great degree of livestock biodiversity which can be elucidated in hosting more than 200 breeds across all the species. Along with producing A2 milk, desi cattle are renowned for withstanding heat, illness resistance, immunity to ticks and parasites and low maintenance requirements. The various indigenous breeds of agricultural animals are mostly the consequence of evolutionary processes. They were improved through field progeny testing & selection. Conservation and propagation of most precious high producing elite buffalo germplasm though the cloning technique is the contribution of the scientists to the buffalo rearing farmers of the country. The Indian Jamunapari goat is one of the ancestors of the American Nubian. Goat milk is an excellent source of vitamin A, is high in digestible protein, contains significantly lower levels of alpha-S1-casein, contains slightly less lactose, and has more non-digestible carbohydrates than cows' milk. Besides, the scientists identified new biopeptides in goat milk having human health benefits. The scientists focused on more lambs per lambing, and developed & propagated Avishan sheep with high fecundity (2-4 lambs/lambing) for higher mutton production. The sheep husbandry sector contributes nearly 8% of total meat production, and employs nearly 6 million people in the country.

The eastern states are amongst the most under-developed states of India characterized by high population density, poor infrastructure, dominance of small sized land holdings, low levels of per capita income and very high incidence of rural distress. Poverty persists in the region because of limited and inequitable access to productive resources, such as land, water, improved inputs and technologies, easy credit, as well as vulnerability to drought and other natural disasters. Livestock comprises an important component of the rural economy in the reason. The total livestock population in major eastern Indian states (in million no) are Assam (10.09), Bihar (36.54), Chhattisgarh (23.61), Odisha (18.17) & West Bengal (37.48). Livestock contribute 38% to the gross value of agricultural output in Bihar and 20% in Odisha. In the past one decade ending 2009-10, livestock sector in both the states has grown at a rate of around 8% outperforming most other states in the country. In fact, in these states the estimated growth has been almost twice the rate of growth in livestock sector at all-India level. Growth in livestock sector in India and elsewhere has been proven to be more pro-poor than the growth in crop sector because of the concentration of livestock at the lower-end of land distribution. In Bihar as well as Odisha also, livestock production is largely in the domain of small landholders who are often poor; hence the poor farmers are expected to benefit more from the expanding livestock sector.

Odisha is an important constituent and southernmost state of eastern India that comprises Assam, West Bengal, Bihar, Jharkhand, Chhattisgarh and 27 eastern districts of Uttar Pradesh. This component of eastern India is located between the parallels of 17.49'N and 22.34'N latitudes and meridians of 81.27'E and 87.29'E longitudes. The geographical area of Odisha in the eastern India is blessed with abundant natural resources and has along coastline, which is 480 km long and 10-100 km wide that forms a part of east coast of India, the coastal territory is drained by a number of rivers like Mahanadi, Brahmani, Baitarani, Devi, Budha Balanga, Subarnarekha, Rushikulya and some other smaller ones. Odisha is a repository for one-fifth of India's coal, one-fourth of iron ore, one third of bauxite and abundant quantities of chromite. Major aluminium and steel plants are located in Odisha. It takes the pride as the first state in India to initiate privatization of its electricity transmission and distribution. Around 60 per cent of its population depend on agriculture. As per 2011 Census, the percentage share of cultivators and agricultural workers in the state has declined from about 65 per cent in 2001 to 62 per cent. Earlier, Odisha had high deficit levels as close to 50 per cent to GSDP in 2013-14, which has now declined to less than 20 per cent of GSDP. Thus, the overall macro scenario is rapidly improving in Odisha. The advent of the Green Revolution coupled with an innovative agricultural extension system enabling effective dissemination of the research outcomes, made a paradigm shift in Indian Agriculture. Although the technical knowhow associated with the green revolution reached Odisha a little late but nevertheless consistent efforts were made by all concerned to make-up for the delay in harnessing its benefits. The key factors of green revolution i.e., use of high yielding varieties of seeds, increased use of chemical fertilizers and pesticides are directly linked to the creation of irrigation facilities. Probably, the lack of such facilities in Odisha became a major hurdle for the spread of the green revolution related to scientific farm management practices. The frequent occurrence of natural calamities like drought, flood and cyclone was another impediment in taking forward the agriculture from subsistence level to the present status of commercial agriculture. The progress in green revolution was definitely slow in initial years as the entire process was dependent on the farmer's educational level and adaptability to new technologies, availability of requisite inputs and information, and non-entrepreneurial skill, poor connectivity, warehousing facilities / lack of agro-based industries etc., which also acted as deterrents for the growth of agriculture. However, the things started moving in a positive direction at a faster pace with the introduction of State Agriculture Policies and focused importance of the government to this sector. The agriculture sector in Odisha provides employment and sustenance to more than 60 percent of the population. Cereals constitute more than 90 per cent of total food grains production, and paddy continues to be a dominant crop. However, there has been a gradual shift from paddy to cash crops and from local variety of paddy to HYVs. The state has more than 35 per cent area under irrigation. The performance of the agriculture sector is highly dependent on the impacts of natural shocks in the form of cyclones, droughts and floods, resulting in wide fluctuations in the annual growth of this sector. The share of agricultural products to GSDP has been going down over the years. Moreover, there are fluctuations in agricultural income in the state over the years, triggered by environmental factors. In the current decade, the state economy of Odisha has witnessed a shift from agriculture towards industry and service sectors. Besides these shifts, agriculture is still being considered as a priority sector for the State. Therefore, a new Agricultural Policy has been implemented since 2013. In the drought-hit western part of the state, the current rate of interest on crop loan has been reduced to ensure continuance of farming by the small, medium and marginal farmers of the state.

Livestock production confronts a number of biotic and abiotic constraints including poor quality of animal germplasm, frequent occurrence of diseases and acute feed and fodder scarcity. Besides, many socio-economic and institutional factors such as under-developed veterinary and breeding infrastructure, shrinking common grazing lands and lack of farmers' access to markets, credit and information restrict farmers realizing production potential of their animals. Overcoming these challenges requires evolving suitable development strategies and targeting thereof to the regions and species that have potential to generate significant economic and social gains without causing any damage to the environment and natural resources. The share of livestock sector to Gross Value addition (GVA) during last one decade to that of agriculture and allied sector is oscillating from 17 to 20% (Table 1).

Table 1. Share of Agriculture & Allied and Livestock Sector in GVA (Rs Crore) at Current Prices

Year	GVA (Total)	GVA (Agriculture & Allied)		GVA (Livestock Sector)	
		Amount	% Share to total GVA	Amount	% Share to total GVA
2011-12	81,06,946	15,01,947	18.5	3,27,334	4
2012-13	92,02,692	16,75,107	18.2	3,68,823	4
2013-14	1,03,63,153	19,26,372	18.6	4,22,733	4.1
2014-15	1,15,04,279	20,93,612	18.2	5,10,411	4.4
2015-16	1,25,74,499	22,27,533	17.7	5,82,410	4.6
2016-17	1,39,65,200	25,18,662	18	6,72,611	4.8
2017-18	1,55,05,665	28,29,826	18.3	7,85,683	5.1
2018-19	1,71,75,128	30,29,925	17.6	8,82,009	5.3
2019-20	1,83,55,109	33,58,364	18.3	9,77,730	5.2
2020-21	1,80,57,810	36,09,494	20	11,14,249	6.2

Source: National Accounts Statistics-2022, Central Statistical Organisation, GoI

Livestock Census

Odisha possesses 4 per cent of the total bovine population of the country. According to the 19th Livestock Census, Odisha has witnessed secular decline in bovine population during the inter-census period from 2007 to 2012. During the same period, there has been an increase of 77 per cent in lactating exotic and cross-bred cattle and 3.5 per cent in indigenous cows. The number of lactating buffaloes has fallen by 11 per cent from 1.43 lakh to 1.27 lakh in the 2012 census. In order to give a major push to buffalo breeding, the Odisha Bovine Breeding Policy, 2015 has been announced for genetic improvement of the buffalo population. The key focus of the policy is also to conserve indigenous buffalo breeds that include nationally recognized Chilika, Kalahandi and Manda along with Parlakhemundi, Jerangi, Sambalpur and Kujang populations.

Cattle and buffalo Breeds

The State Government has come out with new "Odisha Bovine Breeding Policy" on 3rd October 2015" to make dairy farming economically viable. The policy aims to increase number of economically productive milch animals, conserve and improve indigenous germplasm (Table 2), increase local adaptability of dairy animals. The population of different indigenous cattle breeds of Eastern India is given in Table 3. The Odisha Livestock Resource Development Society (OLRDS) is the principal implementing agency of the policy. The policy envisages replacing popular low-milk yielding cow breed gradually with high-yielding Tharparkar and Binjharpuri breeds. It calls for genetic up-gradation programme with the introduction of germplasm of Sahiwal, Gir and Tharparkar. Rather than using stray bulls for mating of cows, more scientific and technology-driven artificial insemination programme has been put in place. The target is to cover at least 40 per cent of the non-descript cattle, 70 per cent of the crossbred cattle and 40 per cent of the buffaloes under organised breeding programme. In order to produce good quality stock with higher production potential, the Government plans to use imported and progeny tested exotic semen.

Table 2. Distribution of Cattle Breeds of Eastern India.

Breeds	Breeding Tract	Utility	Distribution
		Odisha	
Binjharpuri	Primarily in Jajpur district and adjoining areas of Bhadrak and Kendrapara districts of Odisha	Milk and Draught (Dual)	Binjharpur, Bari, Sujanpur and Dasarathapur blocks of Jajapur district and parts of Kendrapara and Bhadrak districts
Ghumusari	Bhanjanagar area of Ganjam and parts of Kandhamal districts of Odisha	Draught	Soroda, Aska, Bhanjanagar, Dharakote and Shergarh blocks of Ganjam district and some parts of Kandhamal district
Khariar	Mostly in Nuapada district of Odisha	Draught	Khariar, Komna, Sinapali and Boden blocks of Nuapada district and parts of Kalahandi and Balangir districts.
Motu	Southern part of Malkangiri district and parts of Chhattisgarh and Andhra Pradesh	Draught and Manure	Podia, Motu, Malkangiri blocks of Malkangiri district and some parts of Chhattisgarh and Andhra Pradesh
		Bihar	
Bachaur	Bihar (Sitamarhi, Darbhanga, Madhubani)	Draught	Bachaur Pargana (Araria, Banka, Darbhanga, Katihar, Kishanganj, Lakhisarai, Purba Champaran, Purnia, Saharsa, Supaul districts)
Purnea	Northern Bihar	Draught	Bihar state (Purnia, Araria, Katihar and Kishanganj districts)
Gangatiri	Chiefly Ballia and Ghazipur districts of Uttar Pradesh and Shahabad and Rohtas districts of Bihar		Eastern Uttar Pradesh, Chandauli, Ghazipur and Ballia districts and in Bihar, Bhabhua (Kaimoor), Buxar, Arrah and Chhapra districts
		West Bengal	
Siri	Darjeeling district of West Bengal and East, West, South and North districts of Sikkim	Draught	Sikkim and West Bengal

Table 3. Population of different indigenous cattle breeds of Eastern Indi.

State	Breeds	Pure (No.)	Graded (No.)	Total (No.)	Percentage Share w.r.t total (%)
Odisha	Binjharpuri	69,406	14,443	83,849	0.1
	Ghumusari	35,626	3,261	38,887	<0.1
	Khariar	13,365	11,656	25,021	<0.1
	Motu	2,31,954	1,683	2,33,637	0.2
Bihar	Bachaur	32,15,259	11,30,681	43,45,940	3.1
	Purnea			2,19,000	Not assessed in the 20th census
	Gangatiri	2,43,153	2,71,392	5,14,545	0.4
West Bengal	Siri	15,278	8,789	24,067	<0.1
Buffalo breeds					
Odisha	Kalahandi	18,123	7,041	25,164	
	Chilika	11,010	2,648	13,658	
Bihar	Luit			114951	Not assessed in 20th census

Table 4. Performance of Cattle Breeds of Odisha.

Breed	**Colour**	**Average Adult Body Weight (kg)**		**Lactation Yield**	**Lactation Length**	**Calving Interval**	**Age at First Calving**	**Average Milk Fat**
		Male	Female	(kg)	(days)	(days)	(days)	(%)
Binjharpuri	Mostly White. Brown, black & grey are not uncommon	255	207	915-1,350	273-308	404	1,230	4.3-4.4
Ghumusari	Mainly white & sometimes grey	208	167	450-650	280-320	412	1,496	4.8-4.9
Khariar	Mostly brown sometimes grey	195	156	308-360	270-290	510	1,520	4.9
Motu	Mainly brown (reddish) sometimes Grey, Few animals arewhite in colour	172	137	100-140	155-170	420	1,573	4.9-5.0
Bachaur/ Bhutia	Grey or greyish white	280	220	750	260	435	950	5% (4.8 to 7.1%)
Purnea	Primarily grey followed by red and black	202	176	609.1	260	450	1480	4.22
Gangatiri	Completely white colored	340	235	1050	230	395	1700	4.9
Siri	Black or brown with white patches	454	363	425.8	273	464	1300	4.1
Lakhimi	brown and grey colour	240	185.0	359 (325-375 kg)	359	440	1430	5.3% (4.3-6.3%)
Thutho	Black or brown, sometimes white patches on face and body	320	280	474	900			
Masilum	Black, brown, and mixture of brown, grey and black	456	170					

(*Source:* **Animal Genetic Resources of India (AGRI-IS), National Bureau of Animal Genetic Resources, ICAR).**

Table 5. Population dynamics of bovines

State	Crossbred cows > 2.5yr	Indigenous cows >3 years	Total Cows	Female Buffalo > 3 years	Total Cows & Buffaloes
Bihar	2,134	5,013	7,147	3,670	10,817
Odisha	751	2,443	3,194	152	3,346
West Bengal	1,541	5,731	7,273	193	7,466
Jharkhand	293	3,165	3,458	435	3,893
Assam	333	3,485	3,818	138	3,956

Table 6. Distribution of Buffalo Breeds of Odisha

Breeds	Breeding Tract	Utility	Distribution
Chilika	Puri and Khordha districts of Odisha and adjoining part of Ganjam district	Milk and Draught	Krushna Prasad, Brahamagiri blocks in Puri district and Balugaon block of Khordha district and parts of Ganjam district
Kalahandi	Mostly found in Kalahandi districts of Odisha	Milk, Draught, Manure	Mainly found in Bhawanipatna, Junagarh, Golamunda and Dharmagarh blocks of Kalahandi district and parts of Rayagada district
Manda	Koraput, Malkangiri and Nawarangapur districts of Odisha	Draught, milk and manure	Hill ranges of Eastern Ghats and plateau of Koraput region of Odisha

Table 7. Milk Production in last two decades in thousand tonnes

Year	All India	Bihar	Odisha	West Bengal	Jharkhand
2001-02	84,406	2,664	929	3,515	940
2002-03	86,159	2,869	941	3,600	952
2003-04	88,082	3,180	997	3,686	954
2004-05	92,484	4,743	1,283	3,790	1,330
2005-06	97,066	5,060	1,342	3,891	1,335
2006-07	1,02,580	5,451	1,431	3,983	1,401
2007-08	1,07,934	5,783	1,625	4,087	1,442
2008-09	1,12,183	5,934	1,598	4,176	1,466
2009-10	1,16,425	6,124	1,651	4,300	1,463
2010-11	1,21,848	6,517	1,671	4,471	1,555
2011-12	1,27,904	6,643	1,721	4,672	1,745
2012-13	1,32,431	6,845	1,724	4,859	1,679
2013-14	1,37,685	7,197	1,861	4,906	1,700
2014-15	1,46,314	7,775	1,903	4,961	1,734

2015-16	1,55,491	8,288	1,930	5,038	1,812
2016-17	1,65,404	8,711	2,003	5,183	1,894
2017-18	1,76,347	9,242	2,088	5,389	2,016
2018-19	1,87,749	9,818	2,311	5,607	2,183
2019-20	1,98,440	10,480	2,370	5,869	2,321
2020-21	2,09,960	11,502	2,373	6,165	2,434
2021-22	2,21,064	12,119	2,402	6,414	2,629

Table 8. Per capita milk availability (g/day) in India and Eastern Indian states.

Year	All India	Bihar	Odisha	West Bengal	Jharkhand	Assam
2009-10	273	175	112	133	130	69
2010-11	281	184	113	137	136	71
2011-12	290	175	112	140	145	70
2012-13	299	188	114	145	146	69
2013-14	307	195	122	145	146	69
2014-15	319	193	121	147	137	70
2015-16	333	202	121	146	141	70
2016-17	351	209	125	149	145	71
2017-18	370	218	129	154	152	71
2018-19	390	228	142	160	162	71
2019-20	406	240	144	165	170	73
2020-21	427	260	143	173	176	75
2021-22	444	270	144	179	187	77

Feed Resources

Odisha is one of the few states that has estimated present demand for green fodder of about 312 lakh MT and that of dry fodder at 139 lakh MT and juxtaposed these figures with availability of about 161 lakh MT green fodder and 106 lakh MT dry fodders (Table 10). Accordingly, the activities of eight departmental fodder seed production units have been to provide fodder seeds at subsidized rates, mini kits, perennial roots and slips to the farmers. Under fodder development programme, farmers are encouraged to develop fodder plots in their own land. It also provides chaff cutters and feeding troughs to the farmers. In addition to fodder plots, demonstrations are also being given for Azolla and UTPS pits.

Under technology-knowledge and strategy programme, breeder fodder seeds are produced which get multiplied into certified seeds in the farmers' fields. Beside State Plan, many new schemes have been introduced to augment feed resources in Odisha.

Table 9. Feed Resources - Availability vs. Requirement in Odisha.

Year	Dry Matter ('000 MT)		
	Availability	**Requirement**	**Surplus/ Deficit**
1992	18,007.8	32,475.5	-14,467.7
1997	16,308.5	30,984.7	-14,676.2
2003	12,780.1	31,860.0	-19,079.9
2007	17,754.6	29,288.0	-11,533.4
2008	17,850.2	29,963.1	-12,112.9
2009	17,850.2	31,240.5	-13,390.3
2010	17,031.8	33,127.6	-16,095.8
2011	7,389.9	34,850.7	-27,460.8

Source: Feedbase 2012, National Institute of Animal Nutrition and Physiology, Bengaluru.

Policies and schemes for dairy development in Odisha

The Department of Fisheries & Animal Resources Development Department (FARD) in Odisha came into existence in 1991 after being bifurcated from the erstwhile Forest, Fisheries & Animal Resources Development Department. Within FARD, there is a Directorate for Animal Husbandry & Veterinary Services (AH&VS) which is headed by a director.

The activities of the department are supported by:

i) Odisha Livestock Resource Development Society (OLRDS) was formed and registered in 2000 for spearheading livestock breeding activities and ensuring timely and meaningful implementation of National Project for Cattle & Buffalo Breeding" (NPCBB) in the State of Orissa for breeding activities.

ii) The Odisha Cooperative Milk Producers Federation Ltd (OMFED) was established in the year 1980. The major objectives of the Federation are to carry out activities for promoting production, procurement, processing and marketing of milk and milk products for economic development of the rural dairy farming community.

iii) Utkal Gomangal Samiti (UGS) was established in the year 1936. The aim and objectives of the Samiti is to bring all round development of the livestock through up-gradation of local indigenous stock by providing improved bulls, cows, calves, buffalo bulls and bucks etc. and to propagate different types of fodder cultivation, cattle feed and encourage public for the same.

iv) State Society for Prevention of Cruelty to Animals (SPCA) (59 of 1960 Central Act): The Government of Odisha has enforced the Act in the state in the year 1976 for the wellbeing of animals. The main objective of this act is to generate public consciousness towards kindness and compassion to animals. Besides this, Animal Birth Control Programme (ABC), relief and rescue operation, animal health camp, media programmes, relief and rescue operations are also organised.

Scope for Livestock Husbandry

Subsidiary income for farmers: Livestock is a source of subsidiary income for many families in India. Diversification of income and employment portfolio is crucial for sustainable rural livelihoods. Livestock sector can play an important role in poverty alleviation, income enhancement and risk reduction for poor rural households

Nutritional security: Experts point out that one of the main reasons behind the poor performance of India in the Global Hunger Index (India ranks 107th out of 121 countries on the Global Hunger Index 2022) is the lack of quality proteins, essential amino acids, and micronutrients in the Indian diet. This could be associated with the low consumption of meat and eggs among citizens. The average per capita consumption of meat and eggs in India is around 5.5 kg and 79 eggs per annum. This is several folds lower than the global average of 40 kg meat and more than 200 eggs per annum. Also, studies show that in India, only six per cent of the total calories consumed come from proteins, compared to the global average of 30 per cent. According to FAO, livestock products such as meat, poultry and egg proteins typically has higher biological value, net protein utilisation, Protein Digestibility Corrected Amino Acid Score (PDCAS) and Digestible Indispensable Amino Acid Score (DIAAS) than vegetable sources. Moreover, foods of animal origin provide multiple nutrients simultaneously, whereas vegetarian foods lack several essential amino acids and bio-available nutrients. It is essential that the livestock resources are efficiently utilised to ensure high produce and availability of wholesome meat in India.

Employment and Gender equity: A large number of people in India being less literate and unskilled depend upon agriculture for their livelihoods. It provides employment to about 8.8 % of the population in India. Animal husbandry promotes gender equality. More than 3/4th of the labour demand in livestock production is met by women. The share of women employment in the livestock sector is around 90% in Punjab and Haryana where dairying is a prominent activity and animals are stall-fed.

Protection against disasters: Livestock are the best insurance against the drought, famine and other natural calamities. Majority of the livestock population is concentrated in the marginal and small size of holdings. Further, agricultural productions get valuable organic manure provided by the livestock.

Export potential: India is a major meat exporter in the world. The livestock also contributes to the production of wool, hair, hides, and pelts. Leather is the most important product which has a very high export potential.

Moving banks: Livestock are considered as 'moving banks' because of their potentiality to dispose of during emergencies. They serve as capital and in cases many times it is the only capital resource for landless agricultural labourers they possess. Livestock serve as an asset and in case of emergencies they serve as guarantee for availing loans from the local sources such as money lenders in the villages.

Weed control: Livestock are also used for biological control of bush, plants and weeds.

Cultural: Livestock offer security to the owners and also add to their self-esteem especially when they own prized animals such as pedigreed bulls, dogs and high yielding cows/ buffaloes etc.

Challenges

Low productivity: Improving productivity of farm animals is one of the major challenges. For instance, the average annual milk yield of Indian cattle is 1172 kg which is only about 50% of the global average.

Vulnerable to disease: The frequent outbreak of diseases like Food and Mouth Disease (FMD), Black Quarter (BQ), Influenza etc. continues to affect livestock health and lowers the productivity.

Environmental concerns

Conversion of forest land: The conversion of forests into agricultural land and livestock ranches is one of the major causes of deforestation globally.

Greenhouse gas emissions: Raising livestock generates 14.5 per cent of global greenhouse gas emissions. The huge population of ruminants in India contributes to greenhouse gases emission adding to global warming.

Limited Artificial Insemination services: Limited Artificial Insemination services owing to a deficiency in quality germplasm, infrastructure and technical manpower coupled with poor conception rate following artificial insemination have been the major impediments. After more than three decades of crossbreeding, the crossbred population is only 16.6% in cattle, 21.5% in pigs and 5.2% in sheep.

Skewed focus on cereals: Livestock sector did not receive the policy and financial attention it deserved. The sector received only about 12% of the total public expenditure on agriculture and allied sectors, which is disproportionately lesser than its contribution to agricultural GDP.

Neglected by the financial institutions: The share of livestock in the total agricultural credit has hardly ever exceeded 4% in the total (short-term, medium-term and long-term). The institutional mechanisms to protect animals against risk are not strong enough.

Lack of insurance cover: Currently, only 6% of the animal heads (excluding poultry) are provided insurance cover.

Antibiotic use in livestock sector: India needs to worry about antibiotic usage in growing animals for food, especially poultry, with rising incomes fuelling more demand for meat. Such antibiotic use could contribute to the spread of drug-resistant microbes, which are already a major public health problem.

Lack of technology support: Only about 5% of the farm households in India access information on livestock technology. These indicate an apathetic outreach of the information delivery systems.

Shortage of fodder: Hardly, 5% of the cropped area is utilized to grow fodder. India is deficit in dry fodder by 11%, green fodder by 35% and concentrates feed by 28%. The common grazing lands too have been deteriorating quantitatively and qualitatively.

Poor market access: Access to markets is critical to speed up commercialization of livestock production.

Lack of access to markets may act as a disincentive to farmers to adopt improved technologies and quality inputs. Except for poultry products, and to some extent for milk, markets for livestock and livestock products are underdeveloped, irregular, uncertain, and lack transparency. Further, these are often dominated by informal market intermediaries who exploit the producers.

Poor slaughtering infrastructure: Likewise, slaughtering facilities are too inadequate. About half of the total meat production comes from un-registered, make-shift slaughterhouses. Unhygienic conditions of a slaughter houses and wet market are prone to the outbreak of zoonotic disease. For instance, outbreak of Covid-19 pandemic has been blamed on the unhygienic conditions of a wet market in China

High logistics cost: Marketing and transaction costs of livestock products are high taking 15-20% of the sale price.

Shortage of grazing lands: Due to industrialization and urbanization, majority of grazing lands are either degraded or encroached.

Other major challenges faced by the sector are inadequate availability of credit, poor accesses to organized markets depriving farmers of proper milk price, and limited availability of quality breeding bulls, Deficiency of vaccines and vaccination set-up, diversion of feed and fodder ingredients for industrial use. Animal Husbandry is a state subject. Hence, it lacks uniformity in implementation of measures to promote livestock sector.

Government Initiatives

Disease Control

National Animal Disease Control Programme (NADCP): The scheme was launched in 2019 for control of Foot & Mouth Disease and Brucellosis by vaccinating 100% cattle, buffalo, sheep, goat and pig population. The overall aim of the scheme is to control FMD by 2025 with vaccination and its eventual eradication by 2030. This will result in increased domestic production and ultimately in increased exports of milk and livestock products.

Intensive Brucellosis Control programme in animals is envisaged for controlling Brucellosis which will result in effective management of the disease, in both animals and in humans. It is a Central Sector Scheme where 100% of funds shall be provided by the Central Government to the States / UTs.

Livestock Health and Disease Control' scheme: It is a Centrally Sponsored Scheme launched by Department of Animal Husbandry, Dairying and Fisheries (DADF) under the Ministry of Agriculture and Farmers' Welfare implemented since August 2010. It aims for prevention, control and containment of animal diseases of economic importance such as Foot and Mouth Disease (FMD), Brucellosis, Anthrax, Black Quarter (BQ), Classical Swine Fever, New Castle Disease (Ranikhet), Avian Influenza (AI), etc.

National Animal Disease Reporting System (NADRS): It is an on-line system of animal disease reporting linking each Block, District and State Headquarters to the Central Disease Reporting and Monitoring Unit in New Delhi.

Professional Efficiency Development (PED): Under this, assistance is given to the State Veterinary Councils and the Veterinary Council of India (VCI) to carry out their statutory functions under the Indian Veterinary Council Act, 1984 as well as to carry out Continuous Veterinary Education (CVE) for in-service veterinarians

Animal health institutes

Central/ Regional Disease Diagnostic Laboratories: In order to provide referral services in addition to the existing disease diagnostic laboratories in the States, a Central and five Regional Disease Diagnostic Laboratories have been set up by strengthening the existing facilities.

Animal Quarantine and Certification Service: The objective of this service is to prevent ingress of exotic livestock diseases into India by regulating the import of livestock and livestock products and for providing export certification of International Standards for livestock & livestock products. There are six quarantine stations in the country.

National Livestock Mission (NLM): It seeks to ensure quantitative and qualitative improvement in livestock production systems and capacity building of all stakeholders. The scheme is being implemented as a sub scheme of White Revolution – Rashtriya Pashudhan Vikas Yojana since April 2019. There are four sub-missions under National Livestock Mission: Those are Sub-Mission on Fodder and Feed Development; Sub-Mission on Livestock Development; Sub-Mission on Pig Development in North-Eastern Region and Sub-Mission on Skill Development, Technology Transfer and Extension.

Fund for Infrastructure Development

Animal Husbandry Infrastructure Development Fund (AHIDF): It would facilitate the much-needed incentives for the investments in the establishment of infrastructure for dairy and meat processing and value addition infrastructure and establishment of animal feed plant in the private sector

Dairy Processing & Infrastructure Development Fund: It aims to provide loan assistance to Eligible End Borrowers (EEBs) such as the State Dairy Federations, District Milk Unions, etc. to modernize the milk processing plants and machinery and to create additional infrastructure for processing more milk.

Rashtriya Gokul Mission (RGM)

It aims for development and conservation of indigenous breeds through selective breeding in the breeding tract and genetic upgradation of nondescript bovine population. Initiatives carried out under the mission includes Gopal Ratna awards and Kamdhenu awards for encouraging farmers/breeder societies to rear Indigenous breeds of Bovines.

Establishment of integrated cattle development centres 'Gokul Grams' and National Kamdhenu Breeding Centres. E-Pashu Haat portal has been created for connecting breeders and farmers. "Pashu Sanjivni" is an Animal Wellness Programme encompassing provision of Animal Health cards ('Nakul Swasthya Patra') along with UID identification.

National Programme for Dairy Development

The scheme aims to enhance quality of milk and milk products and increase share of organized milk procurement. The scheme has two components. The Component 'A' focuses towards creating/strengthening of infrastructure for quality milk testing equipment as well as primary chilling facilities for

State Cooperative Dairy Federations/ SHG run private dairy/ Milk Producer Companies/FPOs etc. The Component 'B' provides financial assistance from Japan International Cooperation Agency (JICA). It aims for the creation of necessary dairy infrastructure for the purpose of providing market linkages for the producers in villages and for strengthening of capacity building of stake-holding institutions from village to State level.

Livestock Census

The Livestock Census started in the country in the year 1919. So far, 20 livestock censuses have been conducted. Similar to population census, primary workers are engaged to undertake house to house enumeration and ascertain the number, age, sex, etc. of livestock/ poultry possessed by every household/ enterprise in rural and urban areas of the country.

Integrated Sample Survey

The production of major livestock products (MLP) namely milk, eggs, meat and wool is estimated on the basis of annual sample surveys conducted under the Central Sector Scheme "Integrated Sample Survey". Surveys are conducted in the entire rural and urban areas of the States and UTs.

Other initiatives

Kisan Credit Cards (KCC) to Livestock Farmers

A special drive has been undertaken by the Department of Animal Husbandry and Dairying for providing 1.5 crore dairy farmers of Milk Cooperatives and Milk Producer Companies with Kisan Credit Cards (KCC).

e-GOPALA

It is web-based application developed by National Dairy Development Board (NDDB). It helps farmers manage their livestock including buying and selling of disease-free germplasm in all forms (semen, embryos, etc). It also informs about availability of quality breeding services and guides farmers for animal nutrition, treatment of animals using appropriate Ayurvedic ethno veterinary medicine.

Interest subvention

Dairy cooperatives and Farmer Producer Organisations engaged in dairy activities are provided with 4% interest subvention for meeting their working capital requirements.

Way Forward

Sufficient resources: Providing sufficient fodder and drinking water is the need of hour to increase productivity of livestock rearing in India, especially in the rain-shadow region.

Promoting Public Private Partnership (PPP): Investments made by Private companies, Cooperative Societies etc. can be used in augmenting veterinary infrastructure.

Marketing: Trade policies like marketing have to be more effective for promotion of various livestock products like egg, fish, milk etc. and providing sufficient price to farmers by reducing influence of middlemen.

Promoting indigenous breeds: Our indigenous breed of cattle shall be promoted, because most foreign breed cattle are not suitable to our climate and even provide low quality produce. There are some exceptions like Jersey cows, but overall introduction of foreign breeds has not been very successful.

Improvement in infrastructure at slaughterhouses and wet markets: Investments need to be increased to improve infrastructure at slaughterhouses and wet markets to enhance food safety and community wellbeing. Creation of live animal markets separately for each species considering bio-security issues cannot be stressed more. Outbreak of Covid-19 pandemic has been blamed on the unhygienic conditions of a wet market in China. There is also a need to develop more hygienic slaughterhouses, waste disposal and effluent treatment facilities in peri-urban areas and to take up training and certification of producers, processors, retailers, butchers etc

Veterinary institutes and Training: The present number of Veterinary colleges/ Universities is required to be increased in order to reduce the shortage of trained manpower in veterinary services. Necessary training and subsidies should be provided to farmers to adopt livestock rearing as an alternate source of income.

Check the abuse of antimicrobials in livestock sector: There should be strict regulatory framework to limit the overuse and abuse of antimicrobials in food animal production.

Research and development: Government shall also focus on Research & Development in livestock sector to increase per livestock productivity to provide more benefits to small and marginal farmers.

Prioritization of Research Agenda: Scarcity of fodder, particularly dry fodder, is one of the largest constraints to animal husbandry in the state. Availability of green fodder is also grossly insufficient to meet the forage requirements of the animal producers. About 40-60% animals were found suffering from inadequate feeding of dry and green fodder for about two to three months in a year. It is, therefore, imperative that any livestock development policy in the state should accord highest priority to ascertaining adequate supply of fodder from all possible sources like fodder imports as well domestic production. Mineral deficiency is another important problem, leading to a number of serious problems like infertility, late calving, infant mortality, etc. This requires development of feeding technologies, cost efficient nutritional supplements or ration as well as intensive training of the framers to tackle this problem. Scarcity of clean and safe water to maintain animal herds is another major problem in Odisha. Often, animals are allowed to drink water from small ponds, drains or other contaminated water bodies that cause frequent gastrointestinal problems and even costs to the lives of the animals. Provision of clean and safe water is, therefore, imperative to avoid losses and enhance productivity of animals, particularly crossbreds. Shrinking pasture and grazing lands is the most important constraint for the development of goat and sheep farming in Odisha. Most of the existing pastures are heavily loaded with vectors like ticks, flees, and lice which are the carriers of disease. For instance, Theileriosis is one of the diseases which is transmitted by ticks and cause serious production loss. It is, therefore, imperative that development of pasture and grazing land should be an integral component of livestock development policy.

Livestock breeding should also receive urgent priority to improve the outreach and efficacy of artificial insemination that requires adequate facilities and trained manpower. Maintenance of proper cool chain to maintain the viability of semen is essential. Still, significant number animals are inseminated by the natural services. The extent of natural services in case of buffaloes and small ruminants are almost cent percent. However, there is lack of sufficient number of certified or progeny tested bulls, bucks or rams. The public R&D efforts should focus on establishment of adequate number of animal breeding farms to improve the efficacy of breeding programs for supply certified progeny tested bulls and pathogen free quality semen.

The state is prone to floods. The coastal regions often witnesses cyclones and even tsunami. These inflict heavy mortality and morbidity losses. There is an urgent need to develop some safe animal shelters in such areas with all requisite facilities for feeding and health.

Conclusion

A silent revolution is taking place in the livestock sector in Eastern India. The sector has been growing at a rate of around 8%. The growth momentum may come under pressure if not supported by technologies, investments, institutions and infrastructure. This study has generated regional and species priorities for development of the livestock sector in these states and also provides a research and development agenda for different livestock species. Amongst various identified constraints, lack of availability of progeny-tested bulls, mineral deficiency, lack of availability of green fodder, foot and mouth disease and infertility rank higher in order of their severity, are needed to be addressed for enhancing livestock productivity.

References

Annual Report 2022-23. Department of Animal Husbandry and Dairying Ministry of Fisheries, Animal Husbandry and Dairying, Government of India

Basic Animal Husbandry Statistics-2022, Department of Animal Husbandry & Dairying, Ministry of Fisheries, Animal Husbandry & Dairying, Government of India.

Dairying In Bihar-A Statistical Profile 2016. National Dairy Development Board.

Dairying In Odisha-A Statistical Profile 2016. National Dairy Development Board.

Dairying In West Bengal-A Statistical Profile 2016. National Dairy Development Board.

H.Pathak, J.P.Mishra, T.Mohapatra(2022), Indian Agriculture after Independence, Indian Council of Agricultural research , New Delhi

Integrated Sample Survey Reports, Directorate of Animal Husbandry and Veterinary Services, Govt. of Odisha.

Odisha Agriculture Statistics (2007-08 to 2013-14), Directorate of Agriculture and Food Production, Govt. of Odisha.

Odisha Economic Survey 2022-23, Planning and Convergence Department, Directorate of Economics and Statistics, Government of Planning and Convergence Department, Directorate of Economics and Statistics. (2023)

Odisha Economic Survey 2022-23. Bhubaneswar: Government of Odisha. Retrieved from https://pc.odisha.gov.in/publication/economic-survey-report

Singh, RKP. Livestock Research and Development Priorities for Bihar and Odisha, International Food Policy Research Institute IFPRI New Delhi, NASC Complex, DPS Road, Pusa, New Delhi, India.

4

Concepts in Cattle Breeding for Sustainable Improvement

Susanta Kumar Dash

College of Veterinary Science & Animal Husbandry, Odisha University of Agriculture and Technology, Bhubaneswar, Odisha, India

1. Introduction

Cattle are one of the important livestock species of the country which forms an integral part of Indian agriculture since ages. Human progress depends on the utilisation of animals and natural resources in a balanced way. Cow, in particular, is considered as a part of the family in India. Cattle have been of the utmost importance for the sustenance of agriculture, nutrition supply and religious practices in India. Frequent references on cattle farming are found in Indian mythological books and historical writings regarding their utilization in agriculture as draught power, milk production and source of manure. Approximately 200 million Indians are involved in livestock farming, including around 100 million dairy farmers. Roughly 80% of bovines in the country are low on productivity and are reared by small and marginal farmers. Most of the local cows are managed on a low input and low output system for production of milk for the household nutritional requirement and sale surplus milk to dairy co-operatives or local milk sellers. Poor genetic makeup of Indian livestock is the main reason for their lower productivity as compared to the recorded potential of animals in temperate climate. With the change in food consumption pattern due to urbanization, increase in health awareness, reduced requirement of draught animal power due to mechanisation, etc., the demand for milk and milk products have increased many folds. To cater the growing demand, there is an immediate need to increase the productivity of the cattle in the country. The Government of India in collaboration with the State Governments implements the breed improvement and developmental Programs such as Rashtriya Gokul Mission (RGM), National Livestock Mission (NLM), National Program for Dairy Development (NPDD) etc., for enhancing the productivity of the Indigenous cattle. The scientific breakthroughs and technology interventions have played key role for development of cattle rearing in a planned way. The Assisted Reproductive Technologies (ARTs) such as Artificial Insemination (AI), Estrus Synchronization, Embryo Transfer Technology (ETT), Ovum Pick-up & In-Vitro Fertilization (OPU-IVF), and Animal Cloning were standardized and perfected for faster multiplication and maximum utilization of superior germplasm(BNTripathi, et al,2022). Intensification of Animal Husbandry and widespread introduction of exotic breeds for cross breeding have brought significant changes in the domestic animal genetic resources scenario. Genetic variability, which was the result of evolution and selection process through quite long years of domestication under specific geographical conditionsand farming systems has been reduced in many locally adapted breeds. Animal husbandry being the state subject, most of the states have framed their respective breeding policy for cattle covering breeding goals looking to the desired genetic improvement, cross breeding strategy, choice of breed and specific breeding programme. As per National Accounts Statistics 2021, the value of output of milk in 2019-20 is Rs 8.4 lakh crores at current prices surpassing total value of output from food grains (Written reply in the Rajya Sabha by Hon'ble Minister for Fisheries, Animalhusbandry & Dairying, posted on 4 February 2022 by PIB, Delhi).

Livestock Scenario in India

1. India is the highest livestock owner of the world.
2. It is a major risk mitigation approach for small and marginal resource poor farmers, particularly across the rain-fed regions of India.
3. It is at the centre of poverty alleviation programs from equity and livelihood standpoints.
4. Livestock productivity has been identified as one of the seven sources of income growth by the Inter-Ministerial Committee under the government's target of doubling farmers' income by the year 2022.
5. India possesses vast diversity of cattle genetic resources spread in various agro-climatic regions. As per the 20th Livestock Census, the total Livestock population is 535.78 million in the country showing an increase of 4.6% over Livestock Census-2012. As per the census, the country possesses 193.47 million cattle of which 51.36 million are crossbred while the rest 142.11 million are Indigenous.
6. Animal husbandry supports the livelihood of almost 55% of the rural population.
7. As per the Economic Survey-2021, the contribution of Livestock in total agriculture and allied sector Gross Value Added (at Constant Prices) has increased from 24.32% (2014-15) to 28.63% (2018-19).

The number of cattle was 155.3 million at the time of independence, which rose to 192.49 million in 2019. On the other hand the annual growth rate was 0.43 percent during 1951-56 which declined to -0.66 percent during 2007-12. During three census years i.e. 1992-97 (-0.56), 1997-03 (-1.18) and 2007-12 (-0.66) annual cattle growth rate was negative. The negative growth rate during last three decades may be attributed to green revolution equipped with high-tech and mechanized form of agriculture which might have reduced the demand of cattle draught power in agriculture. But the total number of cattle in the country is 192.49 million in 2019 showing an increase of 0.8 % over previous census (2012). Only 26.19% of the cattle population is crossbred and rest are indigenous. There is a decline of 6 % in the total Indigenous (both descript and non-descript) cattle population over the previous census. However, people are becoming less interested in rearing cattle due to low productivity of milk, less fat content in the milk, diminishing pastures and lack of labour to manage them.

The genetic resources of dairy animals in India have a wide range of native bovine breeds (53 cattle breeds and 21 buffalo breeds have been registered). Proportion of identified indigenous breeds to the total population in has been reported as 15% and 40% for cattle and buffalo, respectively. In spite of adopting crossbreeding for some decades, only 15-20% of cattle in the country are recorded to be crossbreds. Therefore majority of the animals belong to non-descript category as of now. The animal genetic resources in the country are vast and varied but not properly characterized, evaluated and documented. It is thus necessary to characterize and evaluate various breeds of livestock in their ecosystem. Further decrease in performance is being observed in F2 generations putting a question mark to adaptability of these animals. In contrary, various indigenous breeds which are the result of thousands of years of selection, evolution and development in the process of domestication are best suited to local agro-climatic conditions.

The prospects for a breed depend to a great extent on its present and future function in livestock production systems. As circumstances change, certain breeds are set aside and are faced with the danger of extinction unless action is taken. There are several reasons why the implementation of conservation measures for a particular breed might be considered important. These are genetic uniqueness, high degree of threat, traits of economic or scientific importance (unique functional traits) and ecological, historical or cultural value. The reason for conservation will to some extent, determine the effectiveness of the conservation measures.

It is imperative that the lesser-known varieties and strains of different breeds of domestic livestock be studied in detail and documented. It is only after proper characterization with respect to genetic and phenotypic parameters, the question that whether a population of livestock of any species is having a distinct genetic identity, can well be answered.

Scope and threat in crossbreeding

Crossbreed animals (F1) generated through use of exotic semen on indigenous cattle perform better than the indigenous females with respect to milk yield and reproductive traits like age at sexual maturity in Indian environment. The crossbred animals with 50% exotic inheritance are to be maintained as per policy. So interbreeding among the crossbreds is followed to achieve this. The animals (F2) generated through interbreeding of 1st generation crossbreds do not perform equally as their parents (F1). The decrease in performance with respect to milk yield and reproductive traits as compared to parents and susceptibility to many diseases observed in F2 females has to be tolerated and should be kept in check in future generations without further deterioration. In other words, these future animals should be adaptable to the Indian environment.

In order that the above situation is created, the following steps are recommended to be undertaken.

By crossing the indigenous cattle by the exotic bulls the genes that expressed in temperate climate through generations were introduced to our animals and are made to express in the tropical/ subtropical climate. Milk yield, reproductive characters and other economic traits as well as susceptibility to diseases are controlled by genes having individual small effect (minor genes) but are more in number (polygenes). Such genes are highly dependent on the environment for their expression. Therefore it should be our endeavour to provide conducive environment to these new animals in this soil, which involves improvement of feeding and management of these animals. Availability of good quality ration/ feed and fodder as per requirement should be ensured to realize the full potentiality for performance of these animals in this environment.

Another breeding measure which is needed to be adapted to combact the situation of decrease in adaptability is to practise selection in F2 and in further generations rigorously. This will reduce the depressing effect of gene segregation on the performance of crossbreds of F2 and future generations. Selection of females at field level is difficult as every owner would like to have more progeny from his cows. Selection of males from the same population is therefore needed to be intensified for use as breeding bulls. Bulls from other populations should not be used as this will introduce new genes into the population making more number of loci heterozygous resulting in segregation of genes to a large extent in future generations and decreasing the performance. It is due to breaking of favourable gene combinations that were built in the breeds due to several generations of pure breeding in a specific environment. If selection of the animals within the population is practised rigorously in F2 and future generations, the decrease in performance of crossbreds, which is noticed across later generations can be reduced to a greater extent.

In order that the crossbreds evolve as a new breed and have consistent performance over generations both the above points should be followed and adhered to strictly. If the crossbred animals are allowed to live, reproduce and perform in the existing environment without proper attention as mentioned above there is chance of increase of disease incidence and adaptability problems and slow regression of performance level to the primitive condition in future generations will result in.

With this background the planning should be made accordingly where crossbred animals are involved. These high potential animals should be reared by farmers who are interested in dairy farming through provision of improved feeding and managemental practices to the animals. In low input rearing if balanced ration/ feed and fodder with right type of management in the specific environments and breeding practices can be provided, it may yield good results, otherwise it should be dealt with caution. Under such condition, the F1 crossbreds need to be back crossed with the indigenous breed to regain the unique qualities.

Further, A1 and A2 milk issue puts an edge on native cattle germplasm in developing strategy. A possible link between the A1 genetic mutation and a range of serious health issues, including heart disease, type 1 diabetes, autism, and other aggravating neurological disorders is discussed by different researchers. It is believed that the mutated A1 gene started in the Holstein cow breed hundreds of years ago, but has been passed to many other breeds because the Holstein cow has been used to improve the production of almost all breeds. The health problems are believed to be caused by a tiny protein fragment formed when we digest A1 beta-casein, a mutated protein in the milk. Milk that does not contain this mutated protein is called A2 milk. Today, the majority of the cows in the USA carry the A1 gene and in contrary all the native cattle and buffalo in India carry A2 gene. The A1 gene present in F1 can be bred out of a herd simply by choosing indigenous bulls which are A2/ A2 sires.

More than 95% of the cow milk proteins are constituted by caseins and whey proteins. Among the caseins, beta casein is the second most abundant protein and has excellent nutritional balance of amino acids. Different mutations in bovine beta casein gene have led to 12 genetic variants and out of these A1 and A2 are the most common. The beta-casein proteins found in cow's milk are made of a string of 209 amino acids all linked together. The A1 and A2 variants of beta casein differ at amino acid position 67 with histidine (CAT) in A1 and proline (CCT) in A2 milk, as a result of single nucleotide difference. The proline binds very tightly to the amino acids on either side of it whereas the histidine does not. The histidine breaks off forming a peptide of a string of 7 amino acids called beta- casomorphin-7 (BCM7). Casomorphins have opioid (narcotic) properties and are not digested well by some people. Studies increasingly point to BCM7 as a troublemaker. Numerous recent tests, for example, have shown that blood from people with autism and schizophrenia contains higher-than-average amounts of BCM7. Initial studies on indigenous cow (Zebu type - Bos indicus), buffalo and exotic cows (taurine type - Bos taurus) have revealed that A1 allele is more frequent in exotic cattle while Indian native dairy cows and buffaloes have only A2 allele and hence are a source for safe milk.

Concept in conservation of native germplasm

An alternative to this is to conserve and improve the local indigenous Germplasm in Cattle which have many good qualities like adaptability to low input rearing heat resistance, low incidence of diseases, high fat percentage in milk and less breeding problems. These animals will be suitable to low input conditions. Further draft animal power production is the back bone of present agriculture system, which can well be achieved through this. This will benefit the farmers in many ways. Good quality indigenous cows may be supplied to the people under livelihood programmes for rearing. Only those persons in the area who are interested for rearing high yielding cows by providing improved feeding and management may be encouraged and proper training and exposure be provided to them. Native breeds like Gir, Sahiwal, Red Sindhi and Rathi are classified as dairy purpose breed and have potential for higher quantity of milk production. Besides, cattle breeds like Kankrej, Haryana and Ongole and buffalo breeds like Murrah, Mehsana, Surti, Jaffarabadi, Nili Ravi and Banni are being conserved in-situ by Government and other agencies.

Various indigenous breeds of cattle are the result of thousands of years of selection, evolution and development of the wild species in the process of domestication suiting to the local agro-climatic conditions and meeting the economic utility of the stakeholders. Further survey reveals that though average performance level of these animals is poor the relative variability amongst animals for the traits of interest is often very much higher. This between animal variability can be exploited for higher genetic gains through well planned and executed breeding programmes, which would bring out sustainable improvement in the productivity and production as well. Further the livestock biodiversity needs to be conserved and the breeds with low production level also need attention in this regard.

As we know, the factors changing animal biodiversity are mutation, migration and selection, but change in domestic animal biodiversity mainly depends on following factors.

- Introduction of exotic germplasm
- Restricted use to a few breeds
- Degradation of ecosystem
- Disease and natural disasters
- Fluctuating market requirements

Issues in breeding policy

India possesses the highest number of livestock in the world, but productivity of the animals is the lowest. Low productivity is largely due to poor genetic makeup of the livestock and traditional management and feeding practices followed by the farmers. Let us think regarding the back bone of sustainable improvement viz. breeding policy. Breeding policy is a programme document for genetic improvement of livestock wealth of the area. It depends on livestock wealth, production level and utility, infrastructure and resources available. It should focus on judicious use of input per unit of output for attaining sustainable animal production and management. It has been realized from the past that formulation of ambitious breeding policy does not help achieving goal. It should be implementable and time bound. Due to poor implementation of such policy desired improvement in productivity has not been achieved. The situation in many states is very similar except very few like Gujarat, Maharashtra, Punjab and Haryana.

Issues in genetic improvement services

Genetic improvement in animal in terms of better production, optimum fertility and health and better animal welfare requires meticulous scientific planning. The breed improvement requires setting up genetic selection program through large scale progeny testing, high-quality semen banks that maintain high genetic proven quality bulls whose semen is supplied throughout the State. Sustained genetic improvement can be achieved only through continuous use of bulls of high genetic merit evaluated through progeny testing. Progeny testing is the estimation of breeding value of bulls by studying the average performance of a number of its daughters. Progeny testing is expensive and time consuming but, at the same time it is the most accurate method for estimating the breeding value of the bull, which is half the herd.

This also requires a cadre of AI workers who are trained in door-step insemination services as per the assigned breeding program for the village. The entire system requires continuous monitoring and performance analysis. For undertaking all these tasks, availability of animal records is crucial. Currently, such systems are not available in the State, hence the supply of semen quality in terms of genetic quality is not up to standard. Although the government has been investing lot of money significant improvement has not yet been achieved. If by accessing the animal Id the information on suitable semen can be obtained, then the actual outcome of the breeding policy will be achieved. Of course the AI centres should have the appropriate semen stock at hand. If the present system is continued, expected progress in livestock productivity is difficult to be achieved.

Therefore, breeding policy needs to address the economic utility of the livestock in the farming system and livelihood of farming community and should be based upon the present management and economic profile of the stake holders targeting to sustainable improvement in productivity and production as well.

Conclusion

The reasons for depletion of native breeds includes crossbreeding with exotic breeds, less economic viability, small herd size and large scale mechanization of agricultural operation. The native breeds need to be conserved for genetic insurance in future, scientific study, as a part of our ecosystem, cultural and ethical requirements and for energy sources in future. In general all the native breeds, which have never been inseminated with exotic semen is the real choice to produce the safe A2 milk. Fortunately in the state of Odisha, there are four registered breeds of cattle – Binjharpuri, Ghumusari, Khariar and Motu and three recognized breeds of buffalo – Chilika, Kalahandi and Manda, having potentiality of producing A2 milk under low input system with high disease resistance. Eventually these native unique bovine germplasm were existing till 2010 in spite of awareness and popularization of crossbreeding and artificial insemination since 1962. Eventually the milk fat content of Chilika buffalo was higher than other buffaloes. Out of this total content of poly unsaturated fatty acids (PUFA), the concentrations of n-6 PUFA and n-3 PUFA were 2.74 and 3.25 g/100 g of milk fat, respectively. The ratio of n-3 to n-6 PUFA was calculated to be 1.19 which was quite good which was higher than cattle and comparable with the fish and substantiate the healthiness of human beings. The objective of augmenting A2 milk production can only be achieved if we rear extensively the native breeds with demand of this milk by the consumers, substantiated with the policy of conservation and improvement of domestic livestock resources.

Way forward

- The major thrust should be on genetic upgradation of native cattle using high quality pedigree bulls and by expanding artificial insemination and natural service network to provide services at the farmer's level. Production of progeny tested bulls should be taken up for each native breed.
- Conservation of the livestock should be the National priority and zero tolerance in breeding of cows in the native tract of a breed with another breed has to be implemented. It can be ensured with non supply of other semen to the AI centres in the native tract of a registered breed.
- The crossbreds need to be back crossed with the progeny tested native breed bulls in subsequent generations to regain the unique qualities of the native breeds.
- Development and strengthening of breeder societies, involvement of farmers in breed improvement programmes and market linking of the native cow milk and milk products with banding tag need to be undertaken.

References

BN Tripathi, VK Saxena, AmrishTygi, P.K.Rout & et al, Book Chapter -Achievements in Livestock and Poultry Production in Independent India, Indian Agriculture after Independence Book, ICAR, July 2022

Written reply in the Rajya Sabha by Hon'ble Minister for Fisheries, Animal husbandry & Dairying, posted on 4 February 2022 by PIB, Delhi

5

Promoting Poultry for Eastern India: Emphasis on Backyard Poultry Production in Odisha

S.K. Mishra, B.K. Swain, R.L. Behera and T.S. Thiyagasundaram*

ICAR-Directorate of Poultry Research Regional Station, Bhubaneswar, Odiha, India

**Former Technical Advisor and Head of Poultry Division, Management of Nature Conservation, Department of UAE President's Affairs, Abudhabi, UAE*

1. Introduction

Poultry industry in India stands out as a major success story. It's phenomenal that India's poultry production whichstarted justas a backyard venture before 1960s, has shaped up into a proven agribusiness, where its annual turnover exceeds Rs 30,000 crores. Current global ranking of India is that of number three, in terms of egg production (behind China [1st] and USA [2nd]), and of18th position in Broiler production. Today, Poultry is one of the fastest growing industries of the agricultural sector in India at present. The annual growth of broilers and egg production is 8-10% compared to 1-2% for production of agricultural crops [3].

Poultry production forms an important Agrarian Avenue providing valuable animal proteins for the fast-growing human population of the developing India, with promises of additional income to small and marginal farmers.Rural and backyard poultry farming is very common among landless and marginal farmer families in India and is a rewarding source of additional income. It involves low investment with high returns which can be easily operated by women, children and elder members of the farm families. Poor households generate good income through backyard poultry farming which provides meat and egg rich in protein and energy. Agriculture including animal husbandry provides 30-40% employment in terms of man/woman days in a year. Scarcity of land holdings, land fragmentation and seasonal agriculture aresome of the reasons that crop production only is unable to provide cent percent employment to the rural youths (men and women). In such circumstances, backyard poultry systems using extensive farming provides households with additional income and right to use nutritionally rich food sources. Backyard poultry rearing can get started with minimal investment by a stakeholder, where birds are raised in all natural conditions. It can thus, be considered as an eco-friendly approach capable of meeting manure-needs for farmer's field (Selvam, 2004). Besides this, rural poultry eggs and meat using desi resources are in great demand during festivals and traditional ceremonies (Alders *et al.*, 2003).

Growth and spread of Poultry in different Regions of India

An important aspect of poultry development in India is the significant variation in the industry across different regions. The four southern states - Andhra Pradesh, Karnataka, Kerala and Tamil Nadu - account for about 45 percent of the country's egg production, with a per capita consumption of 57 eggs and 0.5 kg. of broiler meat. The eastern and central regions of India account for about 20 percent of egg production, with a per capita consumption of 18 eggs and 0.13 kg. of broiler meat. The northern and western regions of the country record much higher figures than the eastern and central regions with respect to per capita availability of eggs and broiler meat.

Growth of Poultry across Eastern States

Over the past two decades, the eastern states of India have experienced considerable growth in the poultry industry. Each state's progress is as follows:

West Bengal: West Bengal is a leading poultry producer in the eastern region, witnessing tremendous growth in the backyard poultry sector. The state government's support through schemes and subsidies has contributed to a significant increase in poultry production from 482,000 metric tons in 2000 to 1.2 million metric tons in 2020 [1].

Assam: The poultry industry in Assam has seen growth over the past two decades, with the state government's initiatives supporting farmers in setting up modern poultry farms. Poultry production in Assam increased from 51,000 metric tons in 2000 to 135,000 metric tons in 2020 [1].

Bihar: Bihar has witnessed substantial growth in poultry production over the past two decades, with the government providing support to farmers through various schemes. Poultry production in Bihar increased from 63,000 metric tons in 2000 to 238,000 metric tons in 2020 [1].

Jharkhand: Jharkhand has also experienced growth in the poultry industry over the past 20 years, with the state government promoting poultry farming through subsidies. Poultry production in Jharkhand increased from 21,000 metric tons in 2000 to 76,000 metric tons in 2020 [1].

Chhattisgarh: Chhattisgarh has witnessed significant growth in poultry production over the past 20 years, with the state government supporting farmers through various schemes. Poultry production in Chhattisgarh increased from 49,000 metric tons in 2000 to 155,000 metric tons in 2020 [1].

So, as can be seen from above, most eastern states of India have witnessed reasonable growth in their respective poultry industry, over past 20-30 years. The respective state governments have implemented several schemes to promote poultry farming and encouraged farmers by providing subsidies to for setting up modern poultry farms, which have led to a horizontal spread in production, but also improved the livelihoods of farmers.

Further to above,Let's analyze the growth and development of Poultry in Odisha, where it has made significant strides over the last few decades. Let's look at the pattern of growth in poultry in odisha, in finer details!

Poultry Farming in Odisha

Poultry has been one of the top growing sectors in Odisha providing employment to many people. To promote poultry production in the state, the State Government has given poultry as the *status of Agriculture*. The state government had also formulated an exclusive "**Odisha Poultry Policy 2015**" which targeted to reach production of 100 lakhs of eggs/day by 2020. Presently, the annual egg production has escalated to 307.85 crore during the year 2021-22 from 192.73 crore during 2015-16. The annual growth rate for egg production in the state for the year 2021-22 is highest among all Indian states (26.99%). The present egg production is more than 84 lakhs per day, which was 64 lakhs in 2018-19 and 52 lakhs during 2014-15. The per capita egg availability of the state (67eggs/annum) is however, far away from the National average (~95eggs/annum). According to DAHD data [Basic Animal Husbandry Statistics, (2022)], Odisha ranks 10th among Indian states in egg production.

Odisha stands 13th among Indian states in terms of annual meat production. The poultry sector has made a substantial contribution to meat production in the state. The present poultry annual meat production of the state is 117.07 thousand tonnes which is 54.15% of total annual meat production (216.20 thousand tones). The per capita meat availability is only 4.73kg/annum which has a huge scope for improvement (BAHS, GOI, 2022).

Table 1. Poultry Population Trend in Odisha: Fig in million no

Census	India	% change over previous census	Odisha	% change over previous census	% share of Nation
1992	307.07	---	13.06	---	4.25
1997	347.61	13.2	18.43	41.1	5.30
2003	489.01	40.68	17.6	-4.5	2.79
2007	648.83	32.68	20.6	17.04	3.17
2012	729.21	12.38	19.9	-3.44	2.73
2019	851.81	16.81	27.4	37.68	3.22

In Odisha, the fowl population got hiked by 43%,while the total poultry population increased by 41.1% during 1992-1997 period. The duck and alternate poultry population increased by 11.2% during the period (16th livestock Census, GOI, 1997). The total poultry have increased from 17.61 million numbers in 2003 to 19.89 million numbers in 2012. There is a decrease of 3.44% in the poultry population during the inter-census period (2007-2012). The fowl population had hiked from 16.88 million in 2003 to 19.42 million numbers in 2012. The duck population has decreased from 0.61 million numbers in 2003 to 0.36 million number in 2012. The Turkey and other birds have decreased from 0.11 million in 2003 to 0.10 million in 2012. Hearteningly, the poultry population in the state was hiked by 37.68%, during the inter-census period (2012 to 2019).

Native chicken like Hansli, Vezaguda, Kabri, Khairi, Gujuri, Dumasil, Kalua, Dhinki, Kalahandi, Phulbani and Khadia are popularly reared by the rural communities of Odisha fulfilling their socio-economic and cultural needs (Singh *et al.*, 2002). Indigenous/native Hansli breed chickens are playing a pivotal role in the rural economy of Mayurbhanj and Keonjhar districts of Odisha (Mohaptra *et al.*, 2006).The native Hansli chicken breed was registered by ICAR-NBAGR in 2018. Native ducks like Kuzi, Moti and exotic ducks like Khaki Campbell, White Pekin are popular in the state. Besides, Quail farming is getting popular in few pockets of the state.

Schemes & state Government support to promote poultry farming in the state

The state government has considered commercial poultry farming as an opportunity for agro-based rural industrialization to solve the problem of unemployment and under-employment in rural areas. The state declared poultry as agriculture and took promotional efforts like reducing electricity tariff for poultry farming, supply of eggs in mid-day meal scheme of primary schools. Fiscal incentives for setting poultry processing units under the National Mission for Food Processing Policy have been encouraged to establish marketing chain and food safety.

Schemes for the Commercial land Backyard layer and broiler farming

As per the State Agriculture Policy, 2013 the state government is supporting poultry farmers for establishing new as well as expanding the existing commercial layer units by providing subsidies. Broiler farming can be undertaken either individually or through integrators. Financial assistance (bank loans) from cooperative and commercial banks is available. The capital investment subsidy under poultry Venture Capital Fund (PVCF) of National Livestock Mission through NABARD and Sate Agriculture Policy 2013 is 40% of the fixed capital (land cost excluded), 50% for SC/ ST/Women/ Graduates from Agriculture and allied disciplines limited to 50 lakhs. The technical officers from Fisheries and Animal Husbandry Department, Krishi Vigyan Kendra and Banks will provide guidance for formulation of projects for the entrepreneurs. The organized backyard poultry can serve as a vehicle for rural development by ameliorating poverty and protein hunger. Dual-purpose colored birds like Vanaraja, Giriraja, Chhabro, Black rock, Gramapriya can be procured from Government and Private hatcheries. 20 chicks given initially and further15 and 10 chicks at an interval od 16 weeks with assistance for night shelter for the birds are being provided. Under support of duckling supply from ICAR-Directorate of Poultry Research and state duck farms, small scale and large-scale duck farming is being promoted by the state government.

Agencies and Organizations Promoting Poultry Farming in the state

Odisha State Poultry Producers Co-operative Marketing Federation Ltd (OPOLFED): it is a state level apex poultry cooperative organization functioning at the state capital Bhubaneswar for popularizing and marketing poultry products.

The Agricultural Promotion and Investment Corporation of Odisha Limited (APICOL): This is a Government of Odisha organization for aiding agricultural enterprises in the state of Odisha. APICOL promotes agricultural enterprises in the state by assistance in guidance for project formulation, counseling, enterprise development and assistance in project implementation. Capital Investment Subsidy is provided to the Poultry -entrepreneurs for setting up of Commercial Poultry-Enterprises under Mukhyamantri Krushi UdyogYojana (MKUY) through APICOL .

Several private hatcheries, namely, the Suguna Poultry Ltd., Eastern Hatcheries Ltd., Amrit Hatcheries Ltd. are contributing to promote broiler and layer farming in the state. Transmission of knowledge and new technologies are carried out via Krishi Vigyan Kendras (KVK), University Extension Block Programme (UEBP), Information and Communication Wing, Video Project, Distance Education and Agricultural Technology Information Centre (ATIC) of Odisha University of Agriculture and Technology (OUAT). There are 8 State Poultry farms functioning in the State under Central assistance (Samal, 2020). Assistance to State for Control of Animal Diseases (ASCAD) scheme is for controlling economically important diseases of poultry by immunization. The Odisha Biological Products Institute, Bhubaneswar, produces different life saving bacterial and viral vaccines for controlling important poultry diseases in the state. Animal Diseases Research Institute (ADRI), Phulnakhara, Cuttack is responsible for strengthening disease surveillance.

Advantages of Backyard/Rural Poultry Farming

- Backyard poultry farming is considered as an alternate source of income to protect against crop failure and ensures rural women empowerment
- Due to regular availability of poultry meat and eggs nourishes the children and women by removing malnutrition among them
- It acts as an additional source of income for rural women
- It needs minimal use of land, labor and capital investment with quick return
- Rural poultry farming can be easily handled without need of any special attention
- It can be integrated with other farming practices like fish farming , agriculture crops and horticultural crops with increased farm income due to composite farming
- Compared to intensive poultry farming rural poultry farming is more environment friendly with less pollution per unit area of production
- It helps in recycling of available low cost resources like insects, ants, fallen grains, grasses, kitchen wastes etc in to valuable human food chain by converting in to nutritionally enriched meat and egg
- It improves soil fertility for better crop production
- Provides self employment and strengthens rural empowerment of women
- Helps conserving the biodiversity as the native or indigenous birds are well adapted to the local climate and resistant to diseases. There is high genetic and phenotypic diversity in indigenous chickens which can be utilized as base resource for further improving the productivity of backyard chickens.

Constraints of Backyard Poultry Production System

- Non-availability of improved germplasm: Backyard poultry are reared in the rural villages in remote and disadvantageous areas in terms of improper road/rail communication, interruption in power supply, water supply, inadequate infrastructure for poultry shed etc. Therefore, this region does not attract investment from industry, forcing the villages to depend on Government organizations for supply of chicks. Due to unfavourable weather conditions like extreme rain and cold, timely supply of improved germplasm is difficult which interferes in the productivity of BPPS across the globe in India [24]. Improved germplasm should be tested in the farm conditions before being introduced them in village area. These germplasms should not be introduced in core breeding tracts of native and indigenous poultry in order to maintain the indigenous gene pool intact.

- Lack of technical skill: Rearing of improved backyard poultry needs upgradation of knowledge and skills of rural farmers to manage them scientifically to realize their full genetic potential. Farmers do not have sufficient knowledge for scientific management of poultry [69] and skill to control diseases [70]. The success of Bangladesh model to improve backyard poultry production was attributed to upliftment of skill of farmers before introduction of poultry [71]. On the basis of needs of scientific knowledge, extension program as per requirement of farmers should be introduced to create awareness and to focus on backyard poultry sector [72].
- Disease, Predation and Biosecurity Threats: Backyard poultry represents a majority of stocks reared by rural people with minimum inputs .Birds reared in the rural villages with minimum biosecurity measures are prone to attack by predators i.e. wild cats,dogsand birds and predisposed to disease outbreaks [7,73]. The predators in the backyard poultry production system are dogs, cats (domestic and wild), eagles, hawk and vultures [6,74]. Diseases like Newcastle disease(ND) and highly pathogenic avian influenza (HPAI) are zoonotic in nature and can have fatal consequences for poultry as well as humanbeings [7,17]. It was also reported that ND is the most common cause of mortality in the backyard poultry rearing process which can cause 100 % mortality [75]. It is very difficult to implement total biosecurity measures in the backyard poultry production system, however, knowledge on prevention of disease, vaccination and proper housing can significantly reduce the household losses [7]. Mortality due to disease outbreaks has been a main constraint in backyard poultry farming [76]. Loss of chicks due to predation is a major constraint in backyard poultry production system [77]. Biosecurity measures for backyard poultry rearing in India

has been suggested which includes daily cleaning of utensils with ash, supply of portable drinking water, disposal of carcasses by garden burial, washing of eggs and storage of eggs in a cold temperature maintained by indigenous structure[73].

- Lack of Veterinary health services: In developing countries cold chain facilities and vaccines are not easily available to the farmers rearing backyard poultry due to lack of resources and infrastructures in remote areas [75]. Lack of these facilities adversely affect the farmer's access to information regarding disease outbreaks, need of biosecurity measures and timely availability of medicines and vaccines [75]. These issues can be addressed by forming networks of community animal health workers, where training and information are exchanged between veterinarians and communities regarding vaccination and disease control [75,78]. Involvement of women in skill development and training programme can have a positive impact on disease control, successful and effective vaccination in the backyard poultry production system.
- Lack of access to market: Backyard poultry farming is practiced in rural areas in remote places of the country away from the city and poorly connected to the market. Nevertheless, the backyard poultry products are natural or organic in nature, these product could not fetch premium prices due to lack of access to the market. The meat and eggs from backyard poultry are usually sold in the local market in villages or cities where farmers do not get proper price.
- Introduction of improved germplasm of backyard poultry to the backyard poultry production system, increased the productivity of the system in terms of meat and egg leading to the decline in the market price due to the surplus produce available with every household in rural areas. If market innovations are not adopted, BPPS may face severe competition from productsof commercial poultry like cases happened in Thailand [79]. Hence, projects on improvement in productivity of backyard poultry must include forward market linkage of the producers.
- Backyard Poultry as a source of antimicrobial resistance (AMR): Backyard poultry production systems are generally practiced in resource poor setting which involves zero to low input. In these resource poor setting antibiotics are not easily available, the minimal use of antibiotics in BPPS [17,50].If antimicrobials are available easily in the BPPS, there are high chances of indiscriminate use as farmers are not well aware and veterinary extension services are poor [80-81]. The use of antimicrobials can increase with an increase in the intensification of the

backyard poultry production and subsequent linkage with the market [73]. Similar findings were also reported by other Southasian countries [82]. In countries like Philippines, family operated micro-enterprises potentially promote the risk of AMR and exposure to zoonotic diseases due to close proximity of community members to production animals as they are reared in the nearest environment of human population [83-84]. It is the need of the hour to educate rural poultry farmers regarding use of AMR in BPPS along with strengthening veterinary health services, raising chicks in stress-free environments and minimal use of AMR [50,85].

- Any policy for Promoting backyard/rural small holder poultry production as a means of Nutritional and Livelihood security on a Pan India basisneeds to take into account all of the above factors and consider, how well the current Poultry (of 21st century), fits into different parts of country's demography and how efficiently it caters to country's poultry consumption. The per capita egg availability in rural areas is almost half than that of urban areas [3].
- Rural household poultry production already contributes 70 % of total production in most low income, food deficit countries [11].
- About 70 % of the world's rural poor people are dependent on the livestock as a component of their income or livelihood. In Vietnam, majority of poor people are dependent on the poultry for their food and income [4].
- In developing countries including India, backyard poultry is an important part of rural poultry production [5-6]. Rural poor people can fulfil their food demand as well as can get profit from backyard poultry farming [7].
- Backyard poultry farming requires low input and can be managed easily [8]. Backyard poultry farming has been identified as a important animal husbandry practice for eradication of malnutrition, alleviation of rural employment and minimization of poverty in rural areas[9-10].
- Research in village chickens in different countries has indicated that the genetic potential of village chickens is generally not the major constraint in their production performance [12].
- In India backyard poultry population is about 317 million and has increased by 45 % in the last decades and presently contributes 35 % of the total poultry population (20th livestock Census, Government of India).

- Backyard poultry production system has been considered as an integral part of many rural families and an income generating activity for women in developing countries.Women manage most of the activities of BPPS like feeding, watering, cleaning and selling of chicken and eggs.There is growing evidence to link the role of BPPS in enhancing the food, nutritional and social security of poor households and also in the promotion of gender equality.
- In India, BPPS is classified into traditional backyard system(<20 birds with little or no input), semi-intensive farming (50-200 birds under semi-scavenging conditions), small-scale intensive farming (200 or more birds with improved birds under high input system), and native chicken farming (indigenous birds with a run area and complete ration)[21]. Further BPPS is classified into small extensive scavenging(1-5 adult birds), extensive scavenging (5-50 birds),semi-intensive (50-200 birds), and small scale intensive production(>200 broilers or > 100 layers)[22]. The type of BPPS is based on the availability of germplasm, marketing facilities, accessibility to natural food base resources, food habits of the population etc.[22-23].

Backyard Poultry Production with Improved Germplasm Supplies in India

Introduction of either improved germplasm or by adopting improved management practices, improves the production and productivity of backyard poultry production system[22,24].Improved germplasm resembling the native chicken phenotypically have been developed by enhancing the genetic potential for improved growth and egg production in birds. The improved birds resemble the desi/indigenous chicken with multi-colored plumage, longer shanks, higher productivity, adaptability to varied agroclimatic conditions, and better immunity[21,25]. Improved birds are capable of producing better (with similar flavor and texture of meat to that of indigenous chicken) on a low nutritional regime in backyard production system. Development of improved germplasm suitable for backyard production system has been achieved either through selective breeding in native or indigenous birds or by crossbreeding of indigenous birds with exotic germplasm. The former method is slow but changes in production are permanent without losing the peculiar character of indigenous birds[26]. Further, dissemination of selected birds (for higher growth and egg production) can be done at farmers backyard. Improved and native germplasms are crossed to develop birds with higher productivity by exploiting the heterosis of two breeds (native and improved). This method has been found to be successful in Asia and Africa continents as shorter time is required for evolving improved germplasm[21,24]. Due to segregation of genes, a decrease in productivity can be there in future generation, which led to the situation that local farmers depend on regular supply of birds from outside.

Further, introduction of crossbred poultry poses threats to the existence of native breeds due to the dilution or erosion of native germplasm. Adoption of suitable breeding policy with utmost care, improved dual-purpose poultry breed has played a pivotal role in enhancing the food and nutritional security of low- and middle-income countries[21]. Government of India and Indian Council of Agricultural Research (ICAR) envisages through their breeding policy that introduction of improved species should be avoided in the home tracts of the recognized native breeds which will prevent the erosion [21]. Several improved poultry germplasms have been developed in different developing countries for backyard poultry production system. It has been reported that high yielding poultry varieties resembling native poultry, transformed backyard poultry farming into a highly remunerative farming activity in India[21].Backyard poultry farms and converting themselves into semi-intensive (50-200 birds unit) backyard farms mainly with the assistance of improved birds and prepared feed to meet the increasing demand for poultry meat[23].In hot humid sub-tropicalclimate, the survivability of 95 % was reported in improved dual purpose backyard poultry Vanaraja and Srinidhi[24]. The hen day egg production was 140 and 195 eggs for Vanaraja and Srinidhi birds,respectively. Due to high consumer preference in the urban areas for meat and eggs of birds reared in backyard farming, these products fetch premium price. Backyard poultry farming caters to the need of people in various ways like it creates provision for nutritious meat and eggs, food for special family functions, chickens for traditional ceremonies, pest control,petty cash, utilization of minimum inputs and human attention and less environmental pollution[24].The net income per bird was significantly higher (Rs995.97) in Vanaraja compared to local birds(Rs287.22)[27]. The net return from 20 birds unit was 46.78 % more in Vanaraja than native poultry with a benefit-cost ratio of 2.84 in backyard poultry production system[28]. In India, rural poultry farming is an significant source of income to meet the household expenses and improved dual-purpose poultry can be resoured to improve the traditional free-range poultry production[25]. It has been proposed for identification, selection and introduction of tropically adaptable semi-scavenging dual purpose breeds to improve the productivity of BPPS in Tanzania[29]. Presently, efforts are being made to introduce those dual purpose breeds with higher genetic potential to improve the growth and egg production and adaptability to varied agro-climatic conditions in BPPS[30]. Indigenous poultry production system has low productivity but it has potential to achieve profitable and sustainable production through genetic improvement[31]. Enhanced management and selection strategies and genetic crosses of commercial chickens with native fowl is required to make the indigenous chicken production system sustainable[32]. Dual purpose bird Kuroiler has been successfully evaluated under farm conditions in scavenging management systems in Uganda[33]. Improved dual purpose poultry for backyard namely Kuroiler and Sasso are becoming popular in Tanzania compared to local chickens because of higher productivity in terms of meat and egg[34-35].Kuroiler birds weighed heavier than indigenous chickens under scavenging conditions in rural household in Uganda[36].It has been reported that the net present value, net cash farm income and highest probability of attaining economic return were highest in rearing Sasoo strain, followed by Kuroiler and the local chicken was economically least viable [37]. The performance of Kuroiler and Sasso are different in lowland and highland ecology and therefore it is imperative to know breed performances with respect to agroecological differences while introducing improved poultry breeds to a different agroclimatic zone [30]. Several improved poultry varieties suitable for backyard poultry farming have been developed in India. These include Vanaraja, Gramapriya, Srinidhi, Giriraja,Kuroiler and Rainbow Rooster[21]. The successful rearing of these varieties has been reported by various studies[21,24,30,37-39]

Table 2. Growth and egg production performance of improved backyard poultry germplasm.

Sl. No.	Variety	Type (egg/ meat/ dual)	Plumage	Body weight (kg),20 week	Egg production (Annual)	References
1.	Vanaraja	Dual	Brown,black with black glossy tail feathers	1.6	137 110	[10,21]
2.	Srinidhi	Dual	Multicolored with barred plumage	0.98	202 140-150	[21,40]
3.	Giriraja	Dual	Multicolored	1.4	125	[41]
4.	Kuroiler	Dual	Thick reddish brown and barred feathers	0.95-1.77 1.7	98-115 up to 45 weeks 159	[42-43]
5.	Sasso	Dual	-	1.05-2.11	98-112 up to 45 weeks	[34-35]
6.	Rainbow Rooster	Dual	Brownish red	1.65	163	[42]
7.	Nandanam	Dual		1.5	176	[44]
8.	Pratapdhan	Dual	Brown,white yellow feathers	1.6 1.75	165 159	[40,45]

9.	Narmadanidhi	Dual	Black with white silky feathers	1.4	170	[40]
10.	Kamrupa	Dual	Brown and black	1.3-1.5	118-130	[40]
11.	Jharsim	Dual	Multicolored	1.6-1.8	110-130	[40]
12.	Himasambridhi	Dual	Brown	1.2	140	[40]
13.	Fayomi	Egg	-	1.21 at 26 week	150	[46]
14.	Sonali	Egg	-	1.18	156	[47]
15.	Gramapriya	Egg	Brown	1.78	256	[38]
16.	CARI Nirbheek	Egg	Brown	1.70	167.9	[48]
17.	UP-CARI	Egg	Brown frizzle feathers	1.22-1.30	-	[49]
18.	CARI-Shyama	Egg	Black with silky white feathers	1.1-1.2		[49]

Nutrition and Food Security as a Major Angle in Ushering Backyard Poultry

In addition to income inequality increased cost of nutritionally balanced diets has debarred about 3 billion people from access to healthy diets in the world. Majority of these people live in Asia (1.85 billion) and Africa (1.0 billion). The FAO report [50] presents the first global assessment of food insecurity and malnutrition for 2020 and offers some indication of what might look like by 2030 in a scenario further complicated by the prolonged effects of the covid-19 pandemic. The number of undernourished people in the world continue to rise in 2020. At present, moderate to severe food security affects more than 30% of world population and most of this population lives in low and medium in-come countries.Poverty and income inequality are underlying causes of food security and malnutrition. Income inequality in particular increases the livelihood of food insecurity especially for socially excluded and marginalized groups [50].

Severe energy deficiency has been reported in one-third population in South Asia and half of the population in sub-Saharan people of South Asia and sub-Saharan Africa, mostly obtain their energy from staple foods like cereal grains, grain legumes, starchy roots and tubers and have access to small quantity of low quality protein. The percapita consumption of egg and animal protein in these region is low as compared to the world average[51]. There is a need for transforming the food system that can provide nutritious and affordable food for all in a sustainable way. The food system need to provide sustainable livelihood for people who work within them, in particular for small-scale producers in developing countries. The backyard poultry is being maintained by poor households for their food and nutritional security because of its low maintenance cost. In general, poultry egg and meat are consumed by all people in the world without any religious taboo. Backyard poultry converts kitchen or other agricultural wastes in to quality animal protein for human consumption, which is much needed by poor households in developing countries[51]. Rural poor households usually consume cereals that have less bio-available protein and are deficient in vital vitamin and minerals. Poultry egg and meat are rich in protein and essential amino acids[51] with high biological value of eggs. Eggs and meats are rich source of micronutrients and considered as food for alleviating undernutrition and malnutrition in developing countries.

The cheapest source of high quality animal based food densely packed with essential macro- and micro-nutrients are the meat and eggs of backyard poultry [17]. Poultry meat is a good source of highly digestible proteins, B-vitamins (mainly thiamin, vitamin B6 and pantothenic acid) and minerals (iron,zinc and copper)[52-53]. Eggs are rich source of protein with high biological value. The protein quality of the egg is taken as standard for measuring the quality of all other food proteins. Eggs are one of the few foods considered to be a complete protein, as they contain all essential amino acids, important source of essential unsaturated fatty acids (linoleic), oleic acid as monounsaturated fatty acid, also rich in minerals like iron, phosphorous, trace minerals and fat soluble vitamins like A,D,E and K and many water soluble B-vitamins. The egg is a potent source of vitamin D but low in calcium (with exclusion of egg shell) and devoid of vitamin C [54].

Improved backyard poultry contributed significantly to the food and nutritional security of tribal farmers in hilly areas of northeast India[24]. Backyard poultry contributes either directly or indirectly to the food security of rural households. Backyard poultry does not compete with human for feed, thereby improving the availability of enriched nutritious food to rural poor at a minimum cost [17]. The ease and high availability and affordability of poultry meat and eggs could contribute to enhanced nutrition and food security of rural poor people, particularly in vulnerable climatic condition. The backyard poultry farming has a huge potential in enhancing the availability of egg and chicken meat in the rural and tribal areas, besides generating employment, supplementary income and empowerment of women. This has led to food and nutritional security of rural people through consumption of egg and meat of backyard poultry in India [21]. Animal protein sources, improves the nutritional status and linear growth of chicken [55]. Therefore, increased backyard poultry production with improved germplasm could help improve the productivity from backyard poultry leading to overall improvement in nutritional status of the rural people as poultry products are often the only source of animal protein for resource-poor rural/tribal households [56].

Backyard Poultry Production and Women Empowerment

Backyard poultry production is a valuable enterprise because of its role in enhancing the availability of nutritionally enriched chicken meat and egg in rural and tribal areas, employment generation, supplementary income and empowerment of women[21]. Backyard poultry are considered as starter capital to alleviate poverty as they provide high value food with small cash income [57-58]. Rural backyard poultry production is being recognized as an important component of socio-economic improvement among weaker section of the society; especially landless labour, small and marginal farm women. Socially, developing countries have male dominating family system where major income from agricultural produce is in the hands of male farmer. Women are primarily responsible for the care and management of the birds under backyard poultry systems [59]. Backyard poultry is the only resource which is completely owned and controlled by women from the women from the time of selection of the bird to sale/purchase and control over the income earned from the bird in terms of sale of meat and egg [60-61].

Women empowerment through self-employment and entrepreneurship training in different socio-economic sectors like backyard poultry production results in opportunities for upliftment of socioeconomic status. Microsoft is a useful tool for the empowerment of women for generation of self-employment and supplementary income and relatively new approach to solve women's problems related to finance[62]. The three main indicators of women's empowerment in India. In rural poultry farming, women are mainly the owners of the poultry [63]. Studies on socioeconomic status of women involved in rural poultry production indicated that poultry raising is a very familiar activity among rural women in many developing countries including India[64].Women mainly contribute to the income of small farm families which is an employment opportunity for them[65]. Rural women usually rear the indigenous types of domestic fowl in extensive system of poultry production[66] ,which provides regular income to women using little inputs and the production can be easily managed by women in the household [67]. Several workers reported that rural backyard poultry production cannot contribute more income but this type of farming system improve the skill of most of the poor women and help them in enhancing socio-economic and nutritional status [63,67]. Hence, backyard poultry farming should be promoted by the government and non-government organizations as it increases independent decision making ability and the involvement of women in their family affairs, which enhances the socio-economic development of the rural sector [68].

Expected Outcome of the Rural Backyard Poultry Farming

Provides rural employment to unemployed rural youths (men and women) through backyard poultry farming with supplementary farm income.

Minimize migration of rural unemployed people to urban areas for search of employment to sustain their livelihood

It helps in alleviating protein malnutrition among vulnerable groups in rural areas like pregnant women, feeding mothers and children

Women empowerment and skill development through backyard poultry farming has been a great success in rural areas because women are better adopted to delicate handling and management of rural poultry farming as they are sincere in their work and can take care of other family activities in addition to poultry keeping by their enterprising skills to increase the income of the rural households

Backyard poultry farming helps in providing improved living standard of the rural people due to additional income and it also improves socio-economic development, health and educational standards of children's of rural/tribal women.

It helps in recycling natural food base i.e. waste materials (insects, ants, fallen grains, green grasses, kitchen wastes etc), available in rural backyard in to highly nutritious products i.e egg and meat for human consumption.

Since demand for organic food is increasing day by day in India and other developing countries, the products from rural backyard poultry can be sold at premium prices as the egg and meat from rural poultry are considered as organic products as they are reared in free range and maintained on naturally available feeds

Due to the ability of rural poultry farming to integrate with other agricultural practices like paddy growing, horticultural crop production and fish farming, it generates high income due to integration.Backyard poultry farming helps in improving the soil fertility in the backyard of the farmer which can increase the production of other crops and vegetables which adds to the income of the farmer and enterprise becomes sustainable.

Conclusion

As can be summarized from this detailed analysis of Poultry Growth in the country, *vis a vis* the Eastern Region, especially of Odisha, it stands out as a slower but, linear growth of poultry sector for the state of Odisha. More than the Industrial poultry growth, the state of Odisha has continued to imbibe the expansion of Poultry, as an agricultural revolution that mostly suited the backyard type of poultry rearing. As such, the backyard poultry sector can contribute to poverty reduction in Eastern India because most of the poor and marginal farmers have limited resources and they can adopt backyard poultry farming for their livelihood easily. Besides reduction in poverty, backyard poultry products i.e. meat and egg help in improvement of nutritional status of poor farmers as they provide high quality protein enriched with vitamins and minerals. It can provide employment to the rural small scale and marginal farmers. Rural poultry can play an important role in women empowerment. In particular, backyard poultry production significantly improves livelihood and food security of women, children and disabled alike. So, there remains an ample scope of backyard poultry farming in Odishabecause of huge gap between demand and supply of poultry meat and egg. Growth, production and reproduction performance of desi backyard poultry is low; thus to increase their performance, improved varieties of poultry have to be promoted. Backyard poultry production performance can be improved along with full support of services such as veterinary health services, extension and proper marketing strategies. In advancing the causes of Backyard poultry, its major constraints list out as: inadequate flow of technical knowledge; low production performance; unavailability of superior germplasm; fluctuation and increased feed cost; upcoming veterinary health services and sluggishly-upcoming marketing support. While the state of Odisha continues to make strides in both Industrial and Backayard type poultry movements, the latter continues to spread continuously, matching to the economic and social fabric of the people of the state in General. So, as the state advances through the 3rd decade of the century, there remains the needto introduce, improved varieties of poultry which can suit both the industrial as well as backyard farming poultry sectors, with focus on skill development of farmers; Quality of feeding poultry; housing, management and disease control needs of both category of poultry, through strengthened training and extension programmes.

References

1. A, Balogh KD, Garcia AD. Antimicrobials used in backyard and commercial poultry and swine farms in the phillippines: a qualitative pilot study. Frontier in Veterinary Science. 2020;7:329.
2. Ahlers C, Alders R, Bagnol B, Cambaza AB, Harun M, Mgomezulu R, Msami H, Pym B.
3. Ahuja V, Dhawan M, Punjabi M, Maarse L. Poultry based livelihood of rural poor: Case of Kuroiler in West Bengal. F. South Asia Pro-Poor Livestock Policy programme.New Delhi, India. 2008.
4. Aklilu HA, Udo HMJ, Almekinders CJM, Vander Jijpp AJ. How resource poor households' value and access poultry: village keeping in Tigray, Ethiopia. Agricultural Systems. 2008; 96: 175-183.
5. Alam J. Impact of small holder livestock development project in some selected area of
6. Alam MA, Ali MS, Das NG, Rahman MM. Present status of rearing backyard poultry in

7. Alders R, Dos Anjos F, Bagnol B, Fringe R,Lobo Q, Mata B, Young M. Case Study A: Learning about the control of Newcastle disease with village chicken farmers in Mozambique, in : Conroy, c (ed.) Participatory Livestock Research: A Guide.2005a;153-163 (ITDG Publications, Rugby).
8. Alders R, Dos Anjos F, Bagnol B, Fumo A, Mata B, Young M. Controlling Newcastle disease in village chickens: A training manual. ACIAR Monograph 2nd Edition.2003; No.86:128.
9. Alders R, SpradbrowP.Controlling Newcastle disease in village chickens: A Field manual. ACIAR Monograph. 2001; No.82:112.http://www.aciar.gov.au, http://www.kyeemafoundation.org
10. Alders RG, BagnolB,Young MP. Technically sound and sustainable Newcastle disease
11. Alders RG,PymRAE.Village Poultry: still important to millions, eight thousand years after domestication.World's Poultry Science Journal.2009;65:181-190.
12. Alexander DJ, Bell JG, Alders RG. Technology Review: Newcastle disease with special emphasis on its effect on village chickens.FAO Animal Production and Health Paper,Rome ,FAO.2004;161:63.
13. Andrew R, Makindara, J, Mbaga SH, Alphonce R. Economic viability of newly introduced chicken strains at village level in Tanzania: FARMSIM model simulation approach. Agricultural Systems.2019;176.102655.doi:10.1016/j.agsy.2019.102655.
14. Anthra, Deepika G.The Aseel Poultry.Andhra Pradesh Veterinary Journal.2000;3:18-21. antimicrobials use in livestock sector in three South East Asian Countries (Indonesia, Thailand and Vietnam).Antibiotics.2019;8:33.doi:10.3390/antibiotics 8010033. backyard poultry rearing practices at Bhandara district of Maharashtra (India).
15. Bagnol B, Alders RG, Costa R, Lauchande C, Monteiro J, Msami H, Mgomezulu R,
16. Bahta, S., Negussie, K., Swain, B., Dhawan, M. and Tripathy, G. 2022. The Odisha livestock sector analysis. ILRI Project Report. Nairobi, Kenya: ILRI. Bangladesh. Livestock Research for Rural Development.1997;9:Article#25. Retrieved December 16, 2022, from http://www.lrrd.org/lrrd9/3/bang932.htm.
17. Barroga TRM, Morales RG, Benigno CC, Castro SJM, Caniban MM, Cabullo MFB, Agunos
18. Baruah MS, Raghav CS. Viability and economics of backyard poultry farming in west siang district of arunachalpradesh, India. International Journal of Food, Agriculture and Veterinary Sciences.2017;7:9-14.
19. Basic Animal Husbandry and Statistics. 2022, Department of Animal Husbandry and Dairying, Government of India, New Delhi.
20. Billah SM, Nargis F, Hossain ME, Howlider MAR, Lee SH. Family poultry production and
21. Branckaert KLI, Nelson V, Loibooki M. Transfer of technology in poultry production for developing countries. Proceedings of the XXI World Poultry Congress, Montreal, Canada, 20-24 August.2000.
22. Bruyn J, de Wong JT, Bagnol B, Pengelly B, Alders RG. Family poultry and food and nutrition security.CAB Review.2015;10:1-9.
23. ChaibanC,RobinsonTP,FavreEM,OgolaJ,AkokoJ,GilbertM,VanwambekeSO. Early intensification of backyard poultry systems in the tropics: A case study. Animal;2020;14:2387-2396.
24. Chitra P. Comparative study of Nandanam chicken IV(Rhode White Chicken) and deshi chicken rearing under backyard system in rural areas of Salem district of Tamil Nadu. International Journal of Science and Environment Technology.2019;8:1049-1053.

25. Conan A, Goutard FL, Sorn S, Vong S. Biosecurity measures for backyard poultry in developing countries: a systematic review.BMC Veterinary Research.2012;8:240.
26. Coyne L, Arief R, Benigno C, Giang VN, Huong LQ, Jeamsscripong S. Characterizing
27. Da Silva M,DestaS,StapletonJ.Development of chicken sector in the Tanzanian Livestock Master Plan. Tanzanian Livestock Master Plan Brief 7.2017; October:1-4.
28. Deka P, Borgohain R, Deka B. Status and constraints of backyard poultry farming amongst
29. Deka RJ, Zakir AMM, Kayastha RB. Improvement of rural livelihood through rearing of Chara-Chemballi ducks in Assam.World's Poultry Science Journal.2014;70:397-404.
30. Desha NH, Bhuiyan MSA, Islam F,Bhuiyan AKFH.Non-genetic factors affecting growth performance of indigenous chicken in rural villages. Journal of Tropical Resources and Sustainability Science.2016;4:122-127.
31. Desta TT. Sustainable intensification of indigenous village chicken production system: matching the genotype with the environment. Tropical Animal Health and Production.2021b;53:doi:10.1007/s11250-021-02773-5.
32. Desta TT.Indigenous village chicken production: a tool for poverty alleviation, the empowerment of women, and rural development. Tropical Animal Health and Production.2021a;53:doi:10.1007/s11250-020-02433-0.
33. Dolberg F. Review of household poultry production as a tool in poverty reduction with focus on Bangladesh and India. FAO Pro-Poor Livestock Policy Initiative Working Paper 2003; No.6.
34. EpprechtM,VinhLV,Otte J,Roland-Holst D. Poultry and Poverty in Vietnam.HPAI Research Brief;2007.No.1:1-6.
35. FAO. How to Reduce the Use of Antibiotics in Poultry Production. Available online at https://www.fao.org/3/cb6811en/cb6811en.pdf . 2021;(accessed December 08,2022).
36. FAO. The Role of Poultry in Human Nutrition. Poultry Development and Review. 2013. Rome.
37. FAO.Family Poultry Development-Issues,Opportunities and Constraints.Animal Production and Health working paper .2014;No.12,Rome. for safe food production.Food Control and Biosecurity.2018;16:481-517. gender issue. FAO Animal Production and Health Paper, Rome Italy.1998;142:160.
38. GetisoA,JimmaA,AsratM,KebedeHG,ZelekeB,BirhanuT.Management practices and productive performance of Sasso chickens breed under village production system in SNNPR,Ethiopia. Journal of Biology, Agriculture and Healthcare.2017;7:120-135.
39. Gueye EHF. Women and family poultry production in rural Africa. Developement in Practice.2000;10:98-102.
40. Gulukande E, Alinaitwe J, Mudondo H. Improving livelihoods of the urban poor in Kampala city through Kuroiler chicken production.Proceedings of the Conference on International Research on Food Security, Natural Resource Management and Rural Development, Tropentag, Vienna.
41. Guni FS, Mbaga SH, Katule AM, Goromela EH. Performance evaluation of Kuroiler and Sasso chicken breeds reared under farmer management conditions in highland and lowland areas of Mvomerodistrict,EasternTanzania.Tropical Animal Health and Production.2021;53,245.doi:10.1007/s11250-021-02693-4.
42. Hedman HD, Vasco KA, Zhang I. A review of antimicrobial resistance in poultry farming
43. Hedman HD, Eisenberg JNS, Trueba G, Rivera DLV, Herrera RAZ, Barrazueta JV, improvement of small farmers through family poultry in Bangladesh.International Journal of Business, Management and Social Research.2015;01:60-70.

44. Islam MS, Begum IA, Kausar AKMG, Hossain MR, Kamruzzaman M. Livelihood
45. Islam R, Deka CK, Rahman M, Deka BC, Hussain M, Paul A. Comparative performance of kuroiler, rainbow rooster and indigenous birds under backyard system of rearing in Dhubri district of Assam. The Journal of Rural and Agricultural Research. 2017; 17: 40-43.
46. Kassa B, Tadesse Y, Esatu W, Dessie T. On-farm comparative evaluation of production performance of tropically adapted exotic chicken breeds in western Amhara, Ethiopia. Journal of Applied Poultry Research.2021;30,100194.doi: 10.1016/j.japr.2021.100194.
47. Khadda BS, Lata K, Kumar R, Jadav JK, Singh B, Palod J. Production performance and economics of CARI Nirbheek chicken for backyard farming under semi-arid ecosystem in central Gujarat,India. Indian Journal of Animal Research.2017;51:382-386.
48. Khadda BS, Lata K, Kumar R,Kaushal S, Sharma RK. Performance of Pratapdhan chickens under field condition of semi-arid ecosystem in central Gujarat, India.Indian Journal of Poultry Science.2016;51:79-83.
49. Khan AG.Indigenous breeds, crossbreds and synthetic hybrids with modified genetic and economic profiles for rural family and small scale poultry farming in India. World's Poultry Science Journal.2008;64:405-415.
50. Khandait VN, Gawande SH, LohakareAC,DhengeSA.Adoption level and constraints in
51. KitalyiAJ.Village chicken production systems in rural Africa, household food security and
52. Kumar M, Dahiya SP, Ratwan P. Backyard poultry farming in India: A tool for nutritional
53. Kumaresan A,BujarbaruahKM,PathakKA,Chhetri B, Ahmad SK, Haunsi S. Analysis of a village chicken production system and performance of improved dual purpose chicken under a subtropical hill agro-ecosystem in India. Tropical Animal Health and Production.2008;40:395-402.
54. Lhermie G, Grohn YT, Raboisson D. Addressing antimicrobial resistance: an overview of
55. Mehta R, Nambiar RG, Delgado C, Subramanyam S. Annex II: Livestock industrialization project:Phase II-Policy,technical and environmental determinants and implications of the scaling-up of broiler and egg production in India. IFPRI-FAO Project on livestock Industrialization, Trade and Social-Health-Environment impacts in Developing Countries.2003.
56. MurdochJ.Themicrofinancepromise.JournalofEconomicLiterature.1999;37:1569-1614.
57. Murphy SP, Allen LH. Nutritional importance of animal source foods.Journal of Nutrition.2003;133:3932S-3935S.19:205-210.
58. Na Ranong V. Structural changes in Thailand's Poultry Sector and its Social Implications.A
59. Nielsen H, Roos N, Thilsted SH. The impact of semi-scavenging poultry production on the
60. Ogunlade I, Adebayo SA, Fayeye TR. Scope and common diseases of rural poultry
61. OkitoiLO,OndwasyHO,Obali MP. Gender issues in poultry production in rural household of Western Kenya. Livestock Research for Rural Development. 2007;19:205-210.
62. Padhi MK.Importance of indigenous breeds of chicken for rural economy and their improvements for higher production performance.Scientifica.2016;6:1-9. priority actions to prevent suboptimal antimicrobial use in food-animal production. Frontier in Microbiology.2017;7:2114.doi:10.3389/fmicb.2016.02114. production by rural women in selected villages of Kwara state, Nigeria.International Journal of Poultry Science.2013;12:126-129. publication commissioned by the FAO.AGAI.2007;pp37.

63. Rahman MS, Jang DH, Yu CJ. Poultry industry of Bangladesh:entering a new phase. Korean Journal of Agricultural Sciences.2017;44:272-282.
64. Rajkumar U, Paswan C, Haunsi S, Niranjan M. Evaluation of terminal crosses to assess the suitability of PD-6 line as a male line for Gramapriya chicken variety developed for rural poultry. Indian Journal of Animal Sciences.2018a;88:438-442.
65. Rajkumar U, Rama Rao SV, Chatterjee RN. Improved chicken varieties.ICAR-DPR Publication.2018b;Pp:1-42.
66. Rajkumar U, Rama Rao SV,Sharma RP. Backyard poultry farming-changing the face of rural and tribal livelihoods. Indian Farming.2010;59:20-24.
67. Rajkumar U,Rama Rao SV, Raju MVLN,Chatterjee RN.Backyard poultry farming for sustained production and enhanced nutritional and livelihood security with special reference to India: a review.Tropical Animal Health and Production.2021;53,176. doi:10.1007/s11250-021-02621-6.
68. RamappaBS.Developmentofvarities forruralpoultry.In:SouvenironSustainablePoultry production: Rural and commercial approach.2008;3rdMarch,Hyderabad,IndiaPp:1-6.
69. Ramdas SR. Reclaiming endangered livelihoods: untold stories of indigenous women and backyard poultry. World's Poultry Science Journal.2009;65:241-250.
70. Rath KR, Mandal KD, Panda P.Backyard poultry farming in India: a call for skill upliftment. Research Journal of Recent Sciences.2015;4:1-5.
71. Rehault-Godbert S, Guyot N, Nys Y. The golden egg: nutritional value,bioactivities, and emerging benefits for human.Health.2019;11:684. research agenda for water, sanitation and antimicrobial resistance. Journal of Water and Health.2017;15:175-184. Rodriguez GIG, Krawczyk E, Berrocal VJ, Zhang L. Impacts of small-scale chicken farming activity on antimicrobial resistant Escherichia coli carriage in backyard chickens and children in rural Ecuador. OneHealth.2019;8:100112.doi:10.1016/j.onehlt.2019.100112.
72. Sabareeswaran TA, Alimudeen S, Abrar Basha MHS, Hariharan R. Role of Backyard Poultry in Rural Development. Pashudhan Praharee.2022;10.
73. Samanta I, Joardar SN, Das PK. Biosecurity strategies for backyard poultry: a controlled way
74. Samson L, Endalew B, Tesfa G. Production performance of Fayomi chicken breed under backyard management condition in mid rift valley of Ethiopia.Herald Journal of Agriculture and Food Science Research.2013;2:078-081.
75. Sanka YD,MbagaSH,MutayobaSK,KatuleAM,Goromela SH. Evaluation of growth performance of Sasso and Kuroiler chickens fed three diets at varying levels of supplementation under semi-intensive system of production in Tanzania. Tropical Animal Health and Production.2020;52:3315-3322.
76. Sarwar F,Usman M, Umar S, Hassan M, Rehman A, Rashid A. Some aspects of backyard poultry management practices in rural areas of district Rawalpindi, Pakistan. International Journal of Livestock Research.2015;5:14-20. security and women empowerment.Biological Rhythm Research.2021;52:1476-1491. selected areas of Mymensingh district. Bangladesh Journal of Animal Science.2014;43:30-37.
77. Selvam S. An economic analysis of free range poultry rearing by rural women. Indian Journal of Poultry Science.2004;39:75-77.
78. Sharma J.A new breed: Higly productive chickens help raise Ugandans from poverty researcher at the centre for infectious diseases and vaccinology at AUS's biodesign Insitute.2011;Available online at http://scienceblog.com accessed on 16-12-2022.

79. Sharma J, Xie J, Boggess M, Galukande E, Semambo D, Sharma S.Higher weight gain of Kuroiler chickens than indigenous chickens raised under scavenging conditions by rural households in Uganda. Livestock Research for Rural Development.2015;27: Article #178. Retrieved December 2, 2022, from http://www.lrrd.org/lrrd27/9/shar27178.html
80. Sharma RPand Chatterjee RN.Backyard poultry farming and rural food security.Indian Farming.2009;59:48.
81. Singh M, Islam R, Avaste RK. Socioeconomic impact of vanaraja backyard poultry farming in Sikkim Himalayas. International Journal of Livestock Research.2019;9:243-248.
82. Singh M, Mollier RT, Rajesha G, Ngullie AM, Rajkhowa DJ, Rajkumar U, Paswan C, Chatterjee RN. Backyard poultry with vanaraja and srinidhi: proven technology for doubling the tribal farmer's income in Nagaland.Indian farming. 2018b;68:80-82. SP.Bangladeshi backyard poultry raiser's perceptions and practices related to zoonotic transmission of avian influenza.The Journal of Infection in Developing Countries. 2011;6:156-165.
83. Sultana R, Rimi NA, Azad S, Islam MS, Khan MSU, Gurley ES, Nahar N, Luby
84. Thieme O, Sonaiya EB, Rota A, Alders RG, Saleque MA, De'BesiG.Family Poultry Development-Issues, Opportunities and Constraints.Rome:2014; FAO Animal Production and Health Working Paper 12. tribal community of Jorhat district of Assam. Asian Journal of Animal Science.2013;8:86-91.
85. Vijayananda CO, Sudheer K, Rongsensusang. Nutritional Value of Poultry Egg: An Overview. Poultry Punch.2019;9:1-6.
Wegener P, Wethli E, Young M. Improving Village Chicken Production: A Manual for Field Workers and Trainers. Canberra:Austrailian Centre for International Agricultural Research (ACIAR).2009.
86. Weyuma H, Singh H, Megersa M. Studies on management practices and constraints of backyard chicken production in selected rural areas of Bishoftu.Journal of Veterinary Science and Technology.2015;2015:1-9.
Within low resource settings.Animals.2020;10:1264.doi:10.3390/ani10081264.
87. Wong IT, de Bruyn J, Bagnol B, Grieve H, Li M, Pym R, Alders RG.Small-scale poultry and food security in resource-poor settings:Areview.Global Food Security.2017;15:43-52.
88. WujitsS, Van den Berg HH, Miller J, Abebe I, Sobsey M, Andremont A. Towards a Zandamela A, Young M. Contributing factors for successful vaccination campaigne against Newcastle disease.Livestock Research for Rural Development.2013;25:95-102.

6

Role of Women in Agriculture and Strategies for Gender Mainstreaming

Lipi Das[1], S.K. Mishra[2] and Ankita Sahu[3]

[1&3]ICAR-Central Institute for Women in Agriculture, Bhubaneswar-751003, Odisha

[2]ICAR-Indian Institute of Water Management, Bhubaneswar-751023, Odisha

'Empowering women is a prerequisite for creating a good nation, when women are empowered, society with stability is assured'

- Former President Padma Vibhushan APJ Abdul Kalam

Women Folk: A Precious Resource

India's ancient history ascribes a divine status to women. *Laxmi*, *Durga* and *Saraswati* are the three great goddesses of Prosperity, Power and Wisdom respectively. The ancient scriptures declared that 'God lives where women are worshipped'. However, a multitude of derogatory attributes have been ascribed to women in the post-vedic period. According to Manu, "a woman must never be independent. In childhood, she must be subject to her father, in youth to her husband, and when her lord is dead, to her sons". The passive role of women, unfortunately thus visualized, later guided the future course of action leading to their present status of backwardness in general and neglect in the streams of development and scientific arena in particular.

The term 'gender' was first used by Ann Oakley and others in 1970s to describe the characteristics of men and women which are socially determined in contrast to biological differences. The term 'gender' further explains, a neutral term meaning either men or women or both in a particular context. The goal is to improve the status of disadvantaged class and get rid of socially created and approved discriminations. The green revolution in India changed the face of agriculture but apparently it contributed to two general trends (FAO, 1996) where wealthy have benefited more from technological change in agriculture than the less well off andsecondly men have benefited more than women.

The socially constructed gender roles of men and women interact with their respective biological roles which in turn affect the nutrition status of the entire family and also of each gender. In Indian rural milieu, women by and large have limited access to land, education, information, credit, technology, and also the decision making opportunity. In contrast, their primary responsibility covers mainly the child rearing, doing household chores and performing farm related operations of crops, livestock and others. As a result, they rely on developed social networks that act as an informal safety net for the family in times of crisis. On the other hand, even in formal employment rural women typically command lower remuneration rates than their male counterpart, despite they possess the same level of skills. The three pronged burden i.e., productive, reproductive, and social roles, women have less time to attend to their own needs, leisure related or otherwise including food and nutrition. The relatively poor nutrition level of female in their early life reduces learning potential, increases reproductive and maternal health risks, and lowers the productivity and efficiency. Indicators of the level of agricultural performance on income have established a strong and significant negative relationship with indices of under-nutrition, suggesting that improvement of agricultural productivity can be a powerful tool to reduce under-nutrition across vast majority of the population.

Women and SDGs

Women for the longest time have been considered as the key to the success of Sustainable Development. For Instance, 'Accomplishment of Sustainable Development and Women's equality are linked inextricably (WRI, 1994). The initial works in research and development largely focused on the role of women in the domestic sphere and programs which addressed the women's areas in order to include women into the process of development. According to Eco-Feminists, women possess child bearing qualities in alignment to their natural affinity with the nature. Men do not have such qualities. In this context, the Sustainable Developmental Goals try and seek to bring change in the course of Twenty-First Century by addressing crucial issues such as violence, inequality and poverty of girls and women. One of the aims of the SDGs is to bring an end to the women and girl child discrimination by building on the above mentioned achievements. However, huge inequalities still exist in the labour markets across some regions as equal access to employment and jobs are systematically denied to women. Investing in women empowerment not only results in furthering progress with respect to SDG Goal number 5 but also results in alleviation of poverty and sustainable growth of economy.

Gender equality is a Sustainable Development Goal (SDG) in its own right, and it is expressly in relation to the SDG goals on hunger reduction and extreme poverty. Development practitioners and Agricultural policy makers have a responsibility and duty to make sure that women have the opportunity to fully engage and benefit from the process of agricultural development. Simultaneously, there will be a reduction in the extreme hunger and poverty as a result of gender equality in agriculture. The status and role of women in the field of agriculture and rural areas differ depending on the social background, ethnicity, age and location, and these are undergoing rapid change in various parts across the world. Development practitioners, donors, policymakers require the research and knowledge that represents and deals with the unique challenges and complexity in the experiences of women in order to formulate policies and arrive at decisions about the sector which is considered as gender-sensitive. (Babu, V.S, and Patil, B. 2018).

Women in Agriculture

India's agrarian sector has shown its strength along with the hardships of the COVID actuated lockdowns. The agriculture and the allied activities was the sole bright spot among the slide in GDP performance of other sectors, clocking a growth rate of3.4 per cent at constant prices during 2020-21. The share of the agriculture & allied sector in total Gross Value Added (GVA), however, increased to 20.2 per cent in the year 2020-21 and 18.8 per cent in 2021-22. The sector has got renewed thrust due to various measures on credit, market reforms and food processing under the *Atma Nirbhar Bharat* announcements. With structural changes in agriculture, there is greater scope to broaden the range of activities related to agriculture to improve productivity and make way for sustainable growth (Economic Survey, 2020-21).

Agriculture is the primary source of income for women in India which contributes more than 80 per cent of their livelihood of which 33 per cent belongs to agricultural labour force and rest 48 per cent are self-employed farmers. Women farmers are amongst the major stakeholders of the agriculture sector and play a predominant role in various on-farm and off-farm activities. They have abundant roles and responsibilities in the sector and can play instrumental in transforming the face of agriculture. About 79% of women continue to be engaged in agriculture and allied activities as against only 63% of men. About 65% of economically active women are in agriculture sector. They contribute 45-50% of their time infarm activities. Although, women spend 354 min/day and men spend 36 min/day on household activities but, they have access to only 5% of agricultural extension resources (Source: Gender Reference Manual-2017, ICAR-CIWA).

The agriculture sector in India employs 80 per cent of all economically active women (OXFAM, 2018).In India, farm women comprise of 33 per cent of agricultural labour force and 48 per cent of self-employed farmers. In rural areas nearly, 85 per cent of women are engaged in agriculture but on an average only 13 per cent of them own land. But, women specific agricultural research, extension, policies and capacity building programs are often ignored. The magnitude of women's work participation and contribution in agriculture is often ignored by prevailing socio-cultural taboos which create perceptible gender disparity. The traditional patriarchal customs, norms and taboos have restricted women to a secondary status within the household and workplace which has resulted in a huge gender gap.

Women in India have remained invisible as farmers despite the fact that they are major producers of food in terms of value, volume and hours worked in agriculture and allied activities which are ecologically and economically critical. The invisibility of the women as farmers is due to the fact that women are concentrated outside market related or remunerative work and they are normally engaged in multiple tasks where too many women do too much work. They belong to a contingent of unpaid family workers in family enterprises and in wage employment. More than half of the women enter into labour force before the age of 15 years as unpaid family workers compared to one third of men in the same age category. Besides agriculture, women have major share of work in animal husbandry, dairy, fruit, vegetable and fish marketing. With increased concentration of agricultural tasks in the hands of women, feminization of Agriculture is in the offing.

The productivity of women is severely constrained by the division of their time, their dual and triple burden and their lack of access to essential resources including knowledge. Small farm production is gradually more unappealing to males who too frequently discard agriculture in favour of better remunerated work in other sectors, leaving women to manage living on often degraded land.

Since India is a versatile country, its each region is endowed with richness in agricultural bio-diversity and natural resources. The eastern region of India is gifted with ample natural resources but its potential could not be harnessed adequately in terms of improving agricultural productivity, poverty alleviation and livelihood improvement. There is a large gap between potential and productivity of major agri-horti crops, livestock, fisheries etc. in eastern region. A majority of population (82.95%) lives in rural areas compared to 72.18% of Indian average. The eastern region comprises of 162 million people of Below Poverty Line (BPL). Hence, livelihood improvement is a major concern in the region. The region can play a lead role in ushering the second green revolution in the country for fulfilling the food requirements of the country, through improved management of its natural resources and some key initiatives.

The region is bestowed with the most important natural resources, i.e., water. The average annual rainfall of the eastern region is highly erratic which ranges from 1091 to 2477 mm (Avg.1525 mm). The water ecologies in the region are precious resources for agricultural production including fisheries, horticultural and animal husbandry practices. Water productivity is major issue in the eastern region since average water productivity is very low in most of the eastern states. The eastern region contributes 19.46 mha out of the total forest area of the country (76.95 mha), on an average, the eastern states contribute 31.14 per cent to total livestock population of India. In addition to various issues, the trend of feminization of agriculture is also acute in the region. The state such as Bihar's agriculture sector is highly feminized, with 50.1 per cent of the total workforce engaged in farming activities are women. Hence, despite being a huge contributor to the agriculture sectorin the eastern region, the gender issues, problems and constraints in agriculture is also prevalent which constrains the overall agricultural development in the region. Hence, research priorities need to be re-oriented accordingly so as to address these issues. The present chapter attempts to explore the role, involvement, issues and potential of women farmers as potential wealth of agriculture sector in the eastern region of the country. It is suggested that amongst the diverse researchable issues, gender issues are paramount and critical to address for ensuring inclusive development.

Gender Issues in Agriculture

a) **Women's triple role:** The women play three major roles of reproductive, productive and community management in the society. Hence, time is the major limiting factor for women in farming sector. The responsibilities of family, household chores and various socio-cultural norms restrict mobility of women in and outside the society.

   Women work longer hours and their work is more laborious than male workers. Women's contribution to agriculture-whether it be conventional farming or market driven agriculture-when measured in terms of the number of tasks performed and time spent, is greater than men. The significant contribution is rightly highlighted by a research study conducted in the Indian Himalayas which found that in a one-hectare farm, a pair of bullocks works 1.064 hours, a man works 1,212 hours and a woman contributes a maximum of 3,485 hours in a year (Shiva, 1991).

b) **Women are expected to do more drudgery prone activities:** Not only do women perform more tasks, their work is also more arduous than that undertaken by men. Both transplanting and weeding require women to spend the whole day and work in muddy soil with their hands. Women's work, preferably does not involve implements and is based largely on human energy, it is considered unskilled and hence less productive. On this basis, women are consistently paid lower wages, despite the fact that they work harder and for longer hours as compared to men.

c) **Women's work is rarely recognized:** Women's employment in family farms or business is rarely recognized as economically productive, either by men or women and any income generated from this work is generally controlled by the men. The farm income is solely counted as the hard earned income of the male head of the farm family and he alone is counted as "food winner" or "food provider".

d) **The shift from subsistence to a market economy has a dramatic negative impact on women:** The Green Revolution, which focused on increasing yields of rice and wheat, entailed a shift in inputs from human to technical. Women's participation, knowledge and inputs were underrated, and their role shift from being "primary producers to secondary workers". Where technology has been introduced in the domain where women worked, the farm women have often been displaced by men in those areas. The threshing of grain mainly considered as a female dominated task and with the introduction of automatic grain threshers, which are majorly operated by men-women have lost a significant source of income.

e) **Women have unequal access to resources (land):** Access to assets is the single most urgent need for the upliftment of women in general and farm women in particular. Though the Indian legislation permits equal right of man and woman in property yet, the condition in actual sense is not so. Rural women still do not have rights in land ownership. Due to this limitation they cannot take independent decision on various agricultural aspects.

f) **Limited access to input and credit:** Though women make substantial contributions to agricultural development, their access to the most crucial input "credit" is limited. As the women do not own the lands, the credit generally flows in the name of male members. This problem restricts their access to technology.

g) **Inadequate technical competency:** Though women are involved in almost all agricultural operations, yet they have inadequate technical competency due to their limited exposure to outside world. This has forced them to follow the conventional practices which in turn result in poor work efficiency and drudgery.

h) **Poor existing Research and Extension Systems:** Though several technological breakthrough have been observed in the recent past, the technologies by the researchers are not tailored to the specific needs of the farm women. As a result, most of the agricultural operations are performed manually and in an unskilled manner, which result into greater drudgery on the part of the farmwomen.

i) **Wage discriminations between male and female agricultural workers:** Women specific farm operations are still rated as light and unskilled work and are undervalued. Female labour wages remain just half of those of the male wages for the same work in similar conditions. The discriminating stand followed by minimum wages Act is questionable in this context.

j) **Women face food insecurity and malnutrition:** The increased hardships and scarcity of food among the low income households dependent on female members for food production and labour wages largely lead to malnutrition of the female members. Poverty has feminine face; where family food security in adverse, the females are the first to be affected.

k) **Untapped women potential:** Though women have many inherent capabilities like high determination, sense of responsibility, better managerial ability, yet their potential has not been identified by the extension personnel. This potential could be explored by the extension scientists and be communicated to the researchers for proper molding of traditional technologies with modern ones. This could certainly help in agricultural production on sustainable basis.

l) **Less participation in decision-making:** In many rural areas, cultural and social norms tend to prevent women from actively engaging in the decision-making process. Women's lower status and input into household decisions gives them restricted control and decision-making power over productive resources and income generated from farming activities.

m) **Occupational health hazards:** Women and men's close proximity to crop and animals expose them to various health risks and hazards. Women are basically responsible for handling food for both family consumption and selling purpose. As a result, they tend to have greater exposure to occupational hazards and diseases than men.

Gender Gap in Agriculture

Moving onto Gender Gap, As per World Economic Forum, India despite of achieving high rates of economic growth in recent year was ranked at 135th

position among 146 countries in Global Gender Gap Index in 2022 and is the lowest performer in the world in the health and survival sub-index where it is ranked 146.The gap represents an extremely high level of gender inequality thus, reflecting its poor performance on gender related assessments despite being a large geographical entity with gigantic natural resources.However, India's overall gender gap score has improved from 0.625 in 2021to 0.629 in 2022, which is its seventh-highest score during the last 16 years.

The Global Gender Report 2022, which includes the Gender Gap Index, depicted that it will now take 132 years to reach gender parity, with the gap reducing only by four years since 2021 and the gender gap closed by 68.1%. The gender gap is the difference between women and men as reflected in social, cultural, intellectual,politicaland economic attainments or attitudes. The analysis of gender gap has basically focused on four key parameters: economic participation and opportunity, educational attainment, health and survival, and political empowerment.

Gender biasness continues in multiple ways: women are not considered as farmers in Indian policies which limit them from getting institutional supports of the bank, insurance, cooperatives, and government departments. Unequal landholdings, social barriers relying on traditional methods and lack of extension facilities are some other challenges faced by women. Women plights are never ending in agricultural sector, exploitation by middlemen, low wage given to small labourers, heavy discrimination and violence by male counterparts and the impact of technology have suppressed their voices.The ones whose day start before sunrise and continue after sunset, their voices are trapped in four walls and often go unheard owing to their gender. Their voices need to be heard to realize the dream of a progressive India.

Women contribute a significant share in agriculture but their growth and development have been shackled by multiple challenges faced by them. Increasing gender stereotypes, male-centric technology development, decreasing water availability, natural resource degradation, increased small holdings, climate related challenges, have affected farm women in agriculture and hence needs to be addressed. India, like many other countries of the world, has the onerous responsibility of achieving the Sustainable Development Goals (SDGs) by 2030. Out of the 17 SDGs set out by the United Nations, where Goal 5 states Gender equality & Goal 10 states reducing Inequalities. The gap can be closed and the issues can be resolved through appropriate policies and government programmes. Priorities need to be identified and strengthening of the linkages will enable in boosting socio-economic status of women.

Some of the identified existing Gender Gap includes constraints such as lesser access to productive resources by farm women as compared to men, granting fewer and smaller loans to women farmers, only 12.8 per cent of land holdings despite women being crucial in agricultural process. With respect to access to technologies in agriculture, only 5per cent out of 45 per cent agricultural workforce have access to agriculture extension services. If women had equal access as men in developing countries then the outcome would have been different; yields in individual farms can increase by 20-30 per cent, overall 4 per cent agricultural production will increase and ultimately the number of undernourished people will be decreased by 12-17 per cent.

Strategic Gender Need of Women In Agriculture

i) **Creating a repository of gender disaggregated data and documentation:** Gender disaggregated information in the field of agriculture and allied areas are scanty and scattered. Such information need to be collected, collated, synthesized and published in order to make them easily available to the users.

ii) **Technology testing and refinement:** Research efforts rarely take into account the gender needs, which differ between men and women farmers in adoption of technologies. There are numerous technologies in agriculture system. But their performance and suitability for meeting the needs of women are not established. Therefore, programmes should be taken up to identify the relevant technologies, test in women perspective and refine to make them women friendly.

iii) **System development and management:** In the wake of emerging problems related to sustainability, the focus has shifted to studies on performance of systems as a whole and their components. Situation warrants that studies related to women's livelihood and development should be taken up in system approach as women meet their diverse needs from different sources that are inter-related in a varying way and related to larger system. Studies on systems such as farming systems would help find solutions to some of the immediate problems faced by women. Due attention should be given to socio-cultural parameters while developing systems in agriculture and allied sectors.

iv) **Drudgery assessment and reduction:** Farmwomen suffer a lot of drudgery while performing farming operations and household activities. They also suffers from different health problems which negativelyimpact their working efficiency and family wellbeing. But, data on the extent to which women are affected in the working environment and the affect on their work output are limited. Hence, it is required that studies should be commissioned on drudgery assessment, and suitable drudgery reducing interventions should be identified.

v) **Gender sensitive extension approach:** Due to the multiple role of farmwomen, extent of illiteracy and socio-cultural barriers, the access of farmwomen to extension/information is limited, therefore it is required to design gender sensitive extension approaches/ models and its efficacy for the gender. Different extension modules shall be formulated in the various subject matter areas like integrated farming system, post-harvest technology, integrated pest and nutrient management, poultry and fish farming, home garden and homestead farming and test those modules in tribal and rural areas to make them appropriate.

vi) **Capacity building of researchers and extension functionaries:** Stakeholders in research and extension systems, need appropriate approach to recognize the crucial role of women in agriculture and highlight the areas in which their efficiency of work could be enhanced either by technological intervention in agriculture and allied sectors on important problems or by improving their knowledge and skills for better work output.

vii) **Efficient resource management:** Resources, both natural and household, provide an important base for livelihood of women and their families. The adoption of livelihood pattern by women farmers depends on the resource availability and their access to such resources. The resources can be common property resources such as forest, water bodies, fallow lands etc. and household resources like cultivable lands, ponds, livestock and different assets.In recent years depletion of natural resources has adversely affected the livelihood security of women and increased their burden of family maintenance in the event of male migration to other places. Similarly, lack of adequate resources at household level and poor management of existing resources have made poor in general and women in particular vulnerable to hunger and other kinds of deprivations. Hence, there is need to focus on studies related to women's role in resource conservation and management for sustained flow of benefits.

viii) **Gender mainstreaming:** Gender mainstreaming in research and extension is considered an important step contributing towards gender equity. Hence it is important to make gender an essential component of almost all the research undertaken in NARS.

ix) **Gender Responsive Budgeting:** It is the process of entailing a gender-based assessment of budgets, incorporating a gender perspective at all levels of budgetary process in order to promote gender equality.Gender Budgeting is a important tool for achieving gender mainstreaming so as to ensure that the women are getting maximum benefits of development equal to their male counterparts. It is not an accounting exercise but a process of keeping a gender perspective in policy/ programme formulation, its implementation and review.

Gender Mainstreaming

Gender mainstreaming is the process of integrating the gender equality perspective at all stages and levels of policies, programmes and projects. Women and men have different needs and perceptions, living conditions and surroundings, including unequal access to and control over resources, power, human rights and institutions, including the justice system. Secondly, why we need it? It is anapproach to improve the quality of public policies, programmes and projects, ensuring a more organized allocation of resources. Finally, how gender equality issues need to be mainstreamed?Gender analysis and gender impact assessments are fundamental tools for gender mainstreaming. These tools support the implementation of gender mainstreaming in a realistic manner. The other relevant factors those are equally important to ensure proper gender mainstreaming are political will, commitment to and awareness of gender equality issues, knowledge, resources (including expertise) and availability of information.

Gender Budgetingis a powerful tool in the process of gender mainstreaming. It is used as a tool for conducting a gender-based budgetary assessment, incorporating a gender perspective at all levels of the budgetary process, and restructuring revenues and expenditures in order topromote gender equality. In short, gender budgeting is a strategy and a process with the desired impact of achieving gender equality goals.

The importance of focusing on gender is women represent 48 per cent of total population in country, but they face high disparities in access to and control over services and resources and their needs to be more fulfilled in more strategic ways (gender blindness should be eliminated) by addressing practical and strategic gender needs.

Need For Mainstreaming Women in Agriculture

Globally, both the women and men play vital roles in various activities of agriculture from producing to processing for providing the food we eat. Rural women in particular are responsible for half of the world's food production and produce between 60 and 80 percent of the food in most of the developing countries. Despite their contribution, women farmers are frequently underestimated and overlooked in development strategies. There is a tendency among most administrators and policy makers to see "men as farmers" and women asfarmers' wives and to highlight their 'supportive role" rather than their productive role"(Planning Commission 11th Plan Report).

Therefore, for economically and ecologically sustainable agriculture, the total intellectual and physical participation of both men and women in the modernization of agriculture and allied sectors is absolutely essential.The correct information on men and women's relative access to, and control over resources is crucial in the development of food security strategies. Increased access to and control over resources not only improves the economic status of women, their households and communities, but also creates a multiplier effect for economic growth. Accordingly, development of gender specific database of farm families in agriculture and allied activities*viz*., horticulture, livestock management, postharvest management, and availability of extension services were studied in the 11th plan period (2007-12).

Various empowerment and gender based development programmes have been introduced to empower women. Despite all the good intentions and a number of programs, we are still a long way from making women empowered especially in agriculture, which is an important sector for livelihood and employment of women and sustenance of many others. With far less access to latest and advanced agricultural knowledge and technology, a host of other socio-economic factors have had an adverse impact on the lives of women farmers in recent years.To meet these multiple objectives it is very important to study the existing status and knowledge levels of farm women in farming systems and their techno-socio-economic empowerment status. The need of the hour is to provide multiple sources of income to the farming community to divert them from distress conditions and suicides.It is also proven that Self-Help Groups (SHGs) today play today a major role in making rural women self-reliant. The SHG system has proven to be very appropriate and effective in offering women the possibility of breaking the gender stereotypealongwithself exploitation and isolation and transiting them into an self-sustained entrepreneur.The analysis of these SHGs of agrarian families will provide policy-makers with a tool offering direction and focus for the work of significantly improving the economic, political and social potential of rural women by prioritization of women's empowerment.The women friendly technology dissemination process needs to be handled with careful planning and execution emphasizing women farmers' needs, experiences and available resources. The incorporation of information communication technology (ICT) application in present situation is an important aspect forproviding opportunities to the remote farm women to live in close proximity of the scientific input. Strong linkages need to be established between direct ICT interventions and it should be part of the different programmes on agricultural development. Women play a core role in agriculture, but underperform in terms of productivity largely because of lack of access to resources such as finance, skills training, and information services. Mobile technology could bridge this gap, helping to increase productivity and incomes of rural women and their households.

Government Measures For Upliftment of Women in Agriculture Sector

With the celebration and commemoration of the progressive 75 years of independence of India, there is promulgation of the mission of necessitating "Empowered women- Empowered Nation". To promote women farmers in agriculture, the Indian Govt. has provided several schemes and policies for holistic development of farm women, enabling their socio-economic growth and health security. These government flagship schemes and programmes have been reinforced to improve rural women stature in society by creating livelihood opportunities and employment generation. Various schemes such as Prime Minister's Employment Generation Programme (PMEGP), National Livelihoods Mission, *Deen Dayal Upadhayay Grameen Kaushalya Yojana* (DDU-GKY), *Pradhan Mantri Kaushal Vikas Yojana* (PMKVY), *Beti Bachao Beti Padhao*, *Pradhan Mantri Matru VandanaYojana* I (PMMVY), etc. has made substantial impact in creating gender parity and socio-economic empowerment of rural women in India. Through these government beneficiary schemes now the rural women are availing access to education, productive resources, capacity building, skill development, healthcare facilities and diversified livelihood opportunities for making them independent decision-maker and self-reliant.

The Government has taken a number of measures for upliftment of women in agriculture and allied sector :1. Mahila Kisan Sashaktikaran Pariyojana (MKSP), 2. Support to States Extension Programme for Extension Reforms, 3. Sub Mission on Seed and Planting Material (SMSP), 4. National Food Security Mission (NFSM), 5. National Mission on Oilseeds and Oil Palm (NMOOP),6. Sub-Mission on Agricultural Mechanization (SMAM), 7. National Horticulture Mission(NHM). In this context the only and unique Institute ICAR-Central Institute for Women in Agriculture (ICAR-CIWA) has been completely dedicated for undertaking research on issues affecting women in agriculture and strategies for gender mainstreaming. It has been focusing on participatory action research in different technology based thematic areas involving rural women to test the appropriateness of technologies and their refinement in women perspective. (Source:https://pib.gov.in/newsite/PrintRelease.aspx?relid=148196).

Access To Markets and Marketing Opportunity

Despite contributing significantly in the crop and livestock sector, women frequently have poor access to markets than men, and play a restricted role in the commercialization of farm products. This tendency often arises from poor marketing skills, low levels of literacy and customary practices that prevent women from freely leaving the house premises. Women farmers have the constraints like difficulty in traveling long distances, spending long hours, high marketing cost, loss during marketing process etc. Delay in settlement, offering low prices, are some of the constraints created by middlemen that discourage women producers. There could be different approaches to link women groups to market depending on the situation.

The micro-credit through Self-Help Groups (SHGs) has proved to be a strategic tool for organizing rural women in groups which ultimately promotes savings and thrift habits among them to gain access to institutional credit for their socio-economic upliftment and empowerment. The SHGs continue to engage in conventional stereo-typed, low return activities and the basic livelihood concerns of the rural poor women remain largely un-addressed. Under the scheme 'National Institute of Agricultural Marketing', the provisions have been made to organize capacity building programmes for women in the field of modern marketing system.

Women Decision Making Status in The Household For Involvement in Agriculture

Both men and women contribute significantly to agricultural production yet, their access to agricultural production resources differ. The existing literature shows that in spite of women's major contribution to the agricultural sector, women's access to resources such as land, livestock, farm implements, farm technologies, new farming techniques, markets, credit, information and extension services is limited (Ibnouf, 2011; Lemlem *et al.,* 2011). Women's negligible participation in decision making reflects a typical aspect of gender inequality in agriculture sector. This inequality in access to and control over production resources between men and women, called as gender gap in agriculture, hinders women's productivity, limits their livelihood options and exacerbates financial strain on them. In many rural areas, cultural and social taboos tend to prevent women from active involvement in the decision-making process. Women's lower status and input in the household decisions gives them a restricted control and decision-making power over productive resources and income generated from farming activities.

Involvement of Women to Impact on Agriculture and Livelihood

The gender inequalities hinder development has been well documented by the researchers and policy makers. The productivity, capital accumulation, technological progress and theinstitutional framework of production are all affected by gender inequality. Importance of gender is widely reflecting in recent development strategies, programmes and performance indicators. Increasing number women headed households (Presently 13%), alarming number of farmer suicides, the disenchantment of youth towards farming (about 40% of the men farmers want to quit farming), and male migration to urban areas lead to gender role transformation in agriculture.

The farm women's access to agri-production resources can enhance women's efficiency and productivity, the increased adoption of improved farm technologies canincrease agricultural productivity and improve household food & nutrition security. If women had provided the same access to productive resources as men, they can boost yield by 20-30%, improve overall agricultural output by 2.5-4%, and reduce hunger by 12-17% (FAO, 2011). When women and men are relatively equal, economies tend to grow faster, poverty is reduced significantly and well-being of men and women is enhanced.

Gender Equality, Poverty Reduction and Inclusive Growth

The challenge faced by the policy makers in addressing gender gaps and systemic constraints is the complexity and inter-connectedness of the wide range of policy interventions and multiple paths to reform. Since, gender-differentiated market failures, persistent social norms and institutional constraints often unite to reinforce gender inequalities and make improving gender equality much more complex.When there are multiple constraints, they all need to be addressed." (World Bank, 2011). Proper policy formulation and implementation would result in gender equality, and increase growth and resilience.

Strategic Alliances Creating Research-Development-Policy Interface

Certain factors are required to create a Research-Development-Policy interface such as initiating capacity development programmes, joint learning, application and practice of recently learned techniques, enabling environment for wider application and effectiveness of the learned techniques. Apart from these an interface for women needs to be formulated*viz*., P, Q, R, S, & T where P depicts production, productivity and profitability; Q for quality and quantity, R indicates remunerative and research, S reflects sustainability and supportive and lastly T suggests training, trading and technology.

Concept of Gender Sensitive Entrepreneurship Model Through Institute-Industry-Stakeholders Linkage in Convergence Mode

A refined and women centric entrepreneurship model was developed for farm women at ICAR-CIWA to transform them into agri-preneurs. The model primarily focuses on the process of self-actualization, wherein women farmers are motivated to upgrade themselves into leaders in agriculture, followed by their resource development. The base of the 'Entrepreneurship model' is based upon improving profitability in agriculture through industry linkage which provides an assured market to women farmers. The core of the 'Entrepreneurship model' is build upon 'Institute-Industry-Stakeholder linkage'. The initial step of the model, the self-actualization process (Stage I), emphasizes on boosting the self confidence of farm women, update their knowledge and improve their skill as an innovator, a learner and a leader. In stage II, ICAR-CIWA, NGOs, successful women groups, WFPOs create a gender friendly environment to improve farm women's participation by adopting a 4C's approach*viz.,* Convergence, Coherence, Coordination and Collective action. In stage III, farm women are focusedforcapacity development through skill development programmes by collective efforts of ICAR-CIWA, KVKs, SAUs, NGOs and state line departments engage in and for farm women. In stage IV, women-friendly technologies and recent ongoing farm technologies are popularized among women farmers by cooperative contribution by ICAR-CIWA, state line departments and industries. The final stage (stage V), advocates policy interventions such as govt. programmes, schemes and subsidies for supplementing the women beneficiaries. At this stage the assistance of State line departments, NABARD and other co-operatives, NHB, and APICOL etc. will strengthen the agripreneurship sustenance among farm women. The financial assistance from NABARD and other cooperatives, input supply from LAMPS (Large Area Multi-Purpose Societies) can promote entrepreneurial activities carried out by farm women. Finally the market linkage plays significant role in completing the cycle where farm women as an entrepreneur sell their produce/ value-added produce directly without any middlemen intervention. Market linkages are often linked to big industries, for e.g., Agro-based food processing industries. With strengthened multi-stakeholder linkage, the agri-entrepreneurship promotion and sustenance among farm women can be ensured.

Issues and Areas of Attention - Gaps To Be Addressed

Despite tremendous contribution of rural women in the country's economy, they continue to be overlooked, exploited and even further disadvantaged by many development processes. Women have tackled some of the most laborious tasks, yet their contribution to farming is not fully appreciated. Their work goes undervalued in an economy, which puts premium on marketable work. The National Commission on Self Employed Women (1989) pointed out that although women work for longer hours and contribute substantially to family income, they are not perceived as workers by the data collecting agencies and government. Although women are active in both the cash and subsistence agricultural sectors, and much of their work in producing food for the household consumption is important for food security, it is not counted in statistics.

- To achieve sustainable agricultural production, research programme need to target food crops and small livestock, for building the farming expertise of women who are responsible for growing food.
- Women's research needs must be considered in agricultural research for harnessing their special skills in production and biodiversity for the benefit for family. Reorientation of research and extension is required to incorporate gender perspectives.
- Rural women's access to land, water, education, extension, credit and appropriate technology needs to be promoted. Gender equity has to be ensured in the decision making process and the relevant policies and programmes.
- There is a need for increasing women's participation in decision and policy making at all levels. Participation of women in local, regional and national decision making bodies will promote leadership of women.
- In view of the critical role of women in the agriculture and allied sectors, as producers, concentrated efforts are required to ensure that benefits of training, extension and various programmesshould reach them in proportion to their numbers. The capacity building programmes in social forestry, soil conservation, dairy development and other occupations related to agriculture sector like horticulture, livestock including small animal husbandry, poultry, fisheries etc. are need of the hour to benefit women farmers in the agriculture sector.
- Strengthening the role of women in agriculture, by paying special attention to the addition of economic value to their time and labour, providing the needed support services such as credit and technical support for self-help groups will help them in multiple ways.
- There are gender-related constraints that limit women's access to ICTs for social and economic empowerment. These relate to physical access to infrastructure, social and cultural norms, skills and opportunities, and financial resources.

- Gender can be mainstreamed in agricultural policy and project management by providing training and information networking, by generating gender-disaggregated data and by developing skills in gender analysis. Organizational and professional performance evaluation in agricultural and development sectors can also be includedto focus on gender issues. Effective leadership by both women and men helps organizations integrate gender in all aspects of agricultural and rural development management, including the use of and access to ICTs and their associated processes and products.
- Programmes need to be strengthened to bring about a greater involvement of women in science and technology by ensuring that development projects with scientific and technical inputs involve women fully. Special measures need to be taken for their training in areas where they develop special skills like use of agriculture technologies, communication and information technology. Efforts to develop appropriate technologies suited to women's needs as well as to reduce their drudgery require a special focus too.
- Women need to be involved in the policies and programmes for environment, conservation and restoration. Considering the impact of environmental factors on their livelihoods, women's participation needs to be ensured in the conservation of the environment and control of environmental degradation. The majority of rural women still depend on the locally available non-commercial sources of energy such as crop waste, animal dung, and fuel wood. In order to ensure the efficient use of these energy resources in an environmental friendly manner, women should be involved in non-conventional energy resources by spreading the use of biogas, solar energy, smokeless *chullahs*etc. so as to have a visible impact of these measures in influencing the eco-system and in changing the life styles of the rural women.
- Taking into consideration the high risk of malnutrition and disease that women face at all the three critical stages *viz*., infancy and childhood, adolescent and reproductive phase, careful attention need to be paid for meeting the nutritional needs of women at all stages of the life cycle. Special efforts are required to tackle the problem of macro and micro nutrient deficiencies especially amongst pregnant and lactating women as it leads to various diseases and disabilities. For this, sensitization of rural women regarding nutrition education is very much essential.
- The social dimensions of extreme poverty require action on a range of fronts, designed to empower, through awareness of rights. This includes

literacy in general, legal literacy in particular, and training programmes that will assist women in productive activity. It is imperative that such skill based training should be well matched with market requirements.

- Therefore, efforts to improve the position of poor farmwomen have to focus on them as economic actors within a framework of their other multiple roles, as well as the total socio-political environment. Improving women's economic productivity influence their own status and survival in the family as well in community and their recognition at the wider societal level. Various evidence suggests that improvements in their 'bargaining power' within the household and direct access to income largely reduces women's dependency.

Recommendation For Policy, Research and Extension Services

The above overview leads to the conclusions that women's participation in agricultural production is extensive and significant. Women contribute immensely in crops, livestock and allied sectors for household and family maintenance. The exact nature of each activity is determined by social class, specific crops, region, season and size of land holding.The significant contributionsof women's work in agricultural and rural production processes do not necessarily lead to policies meant for women. When drafting agricultural policies, planners hardly ever take into account the impact on women. Such policies, therefore, often have pessimistic impacts on women. A major disadvantage in policy formulation is the lack of adequate data. The Gender Disaggregated Data (GDD) on women and their activities, mainly in non-farm households, is not available to the planners which act as the major hurdle for inclusion of farm women in all policies and development stage of new developmental plan. In this context, the major recommendationsare:

- Personnel at the various planning and decision-making levels should be trained and oriented in understanding the gender issues.
- Appropriate initiatives should be undertaken to improve all statistics related to rural women. All data collected should be gender disaggregated for developing and implementing women friendly developmental plans.
- To understand the perceived constraints of farm women, in-depth qualitative analysis of women's role in the agricultural and allied sectors, their work contributions and level of participation in various agro-based activities must be undertaken.
- Sincere efforts must be made to include women at policy-making and decision-making levels; women should be provided with opportunities to upgrade their professional abilities and skills.

- For involving women in the primary planning and policy-making body of governments it is essential to constitute a women's cell to ensure that women are included in all sectoral policies and budget allocations.
- Efforts should be made to involve women in project identification and formulation especially in development sectors. Also, impact assessment on gender component should be made mandatory in agricultural projects.
- Special attention should be directed towards developing technology to reduce women's work burden, training should be made available to women in the use of the technology that often displaces them; credit and other facilities should also be provided so that women are in a place to make their own decisions critical to their activities and to their families and communities.

Conclusions

As one of the most populous nations with a large percentage of farm womencontributing in agriculture, it is the need of the hour to focus on women's skill upgradation, women friendly technology development, development of women groups, providing equal access to and control over productive resources, collecting gender-disaggregated data for designing women friendly policies and bottom-up gender sensitization for recognizing women farmers in Indian agriculture. The greatest challenge for India especially when it is on the verge of a second green revolution is to generate educated, trained, self-reliant, self-motivated, innovative, responsible and visionary women farmers who can lead our agriculture in a much dynamic and innovative way.

The conscious planning for gender mainstreaming is required to bring social, cultural and attitudinal changes which not only strive for ending the invisibility of women's work participation to agriculture, but also eliminating the drudgery and fine tunes the lives of millions of working women in India. It is significant to recognize that women's empowerment through technologies can raise their status only through a meaningful stimulation. Therefore, there is need to have the participation of women at every stage *viz.,* decision making, program formulation and implementation.

There is a need to tailor technologies to meet the demands of women agricultural workers and to make them readily available for women to access. It should be ensured that technologies are accessible, income generating, women friendly and drudgery reducing in agro-based activities. The technologies that are increasing work efficiency of women should be popularized through multi-location field trials.

Expanding the sphere of women extension workers is required who can better identify the women's needs and constraints, priorities and opportunities to meet their requirements. Adequate funds need to be provided to women extension functionaries for field activities and its recognition. For providing effective extension service more number of women extension personnel, training programs, infrastructural facilities, allocation of gender budgets, sensitization of extension functionaries, etc. should be made and need based extension models should be developed based on the farm women's needs and perceptions.

Accumulating evidence shows that empowering women is not only important in its own right, but also often highly conducive for improving agricultural productivity, food security, and nutrition. Research and development of agricultural technologies and interventions should begin with an understanding of how men's and women's interests as producers and consumers are different and hence, consequently research work should be oriented to address the needs of both as equal partners. Only gender mainstreaming in agriculture can lead to productivity gains and improve developmental outcomes for the next generation. Let us all make a concerted effort to strengthen the role of women in agriculture today for more productive and sustainable agriculture.

Economic empowerment of rural women is very important and they have to be integrated into the developmental goals of the nation and *Atmanirbhar Bharat*. The focus should be on the role of women in strengthening the economy and making self-reliant India. To achieve this, a gender transformative approach is required to transforms gender inequalities addressing their root causes. Hence, a concerted effort is required for bringing gender equity, promoting entrepreneurship and realization of empowerment with planning for training, research and extension activities for farm women development and federating farm women into business group.Concentration on skill development, articulation of Tradition, Technology, Talent and Trade (4T's) and developing farm women based clear road map for agricultural development in the country are essentialto strengthen the role of women in agriculture today for more productive and sustainable agriculture.

References

Das, L., Mishra, S.K. and Rath, N.C. 2006. Decent work and empowerment of women in agriculture. Kurukshetra, 54(7):21-28.

Economic Survey, 2021. Agriculture & Food management, Ministry of Finance, Government of India.2:230-260. Available athttps://www.indiabudget.gov.in/economicsurvey/.

FAO. 1996.From evolution to revolution in agriculture. Rome.Availableathttps://www.fao.org/3/AC621E/ac621e05.htm.

FAO. 2011. The state of food and agriculture 2010-2011. Rome. Available athttps://www.fao.org/3/i2050e/i2050e00.htm.

Ibnouf, F. O. 2011. Challenges and possibilities for achieving household food security in the Western Sudan region: The role of female farmers. Food Security, 3: 215-231.

Lemlem, A., Ranjitha, P., Genvieve, R., Dirk, H. and Susan, M. 2011. Gender. IPMS, Ethiopia.

National Commission on Self Employed Women. 1989. Shramshakti: Report of the National Commission on Self Employed Women and Women in the Informal Sector. Available athttps://indianculture.gov.in/shramshakti-summary-report-national-commission-self employed-women-and-women-informal-sector.

OXFAM India. 2018. Move over 'Sons of the soil': Why you need to know the female farmers that are revolutionizing agriculture in India.Available athttps://www.oxfamindia.org/women-empowerment-india-farmers.

Shiva, V. 1991. Most farmers in India are women, FAO, New Delhi (India), P.21.

Shivaji, D.A., Sarkar, A., Nayak, J., Shaji, A. and Kishtwaria, J. 2017. Gender Reference Manual, ICAR- Central Institute for Women in Agriculture, Bhubaneswar, pp: 1-48.

World Bank Report. 2011. World Development Report 2011: Conflict, Security, and Development. Available at https://openknowledge.worldbank.org/handle/10986/4389.

World Economic Forum 2021. Global Gender Gap Report 2021. Available athttps://www.weforum.org/reports/global-gender-gap-report-2021.

7

Tools and Approaches in Agricultural Extension for Eastern India

Alok Kumar Sahoo[1], Rabindra Nath Padaria[2], Lalita Mohan Garnayak[3] and Aiswarya Sabu[4]

[1]ICAR-Indian Institute of Maize Research, RMR&SPC, Begusarai, Bihar

[2](Extension), ICAR-Indian Agricultural Research Institute, New Delhi

[3]Director of Research, Central Agricultural University, Imphal, Manipur

[4]ICAR-Central Institute for Research on Buffaloes, Hisar, Haryana

Abstract

Sustainable agricultural development in a nation heavily relies on the efficacy of its agricultural extension strategies, methods, service delivery, and processes. While the green revolution and modern extension services have contributed to a surplus in food production, there remains a significant disparity in agronomic and socio-economic outcomes between North Western and Eastern India. Addressing these gaps requires the adoption of effective research-based extension tools and approaches tailored to regional development needs. Extension approaches encompass the broader strategies for designing agricultural extension programs, while tools serve as the means to achieve these strategies. Eastern India's pluralistic extension system showcases various extension tools and approaches. Key extensions include front-line extension services offered by ICAR and SAUs, KVKs, ATICs, ABI of institutes or SAUs, and ministry-based extension services facilitated by NMAET under the Ministry of Agriculture, Government of India, and state departments focusing on agriculture, horticulture, animal husbandry, and fisheries. Management institutions such as MANAGE, NIPHM, IIPM, EEI, and SAMETI play pivotal roles in designing extension policies, executing programs, and building the capacity of extension professionals. Recent extension reforms have introduced public participatory extension models like ATMA, franchised convergence extensions such as the IARI-Post Office Linkage Extension model, and collaborations with CGIAR institutes like CSISA. Public sector target extension approaches are also in place for specific development targets such as the Complex Driven Rainfed Plan, TSP, SCSP, State Plan (RKVY), regional plans (BGREI), Aspirational District Plan, NFSM, MIDH (NHM), among others. Various public-private partnership extension models have emerged, including government-regulated private extension, government-franchised private extension, government-franchised grassroots-led extension, government-supported Agri-graduates ACABC Extension, and para-professional-led extension services. Additionally, there are privatized extension models like Input Agency Extension, Consultancy Service Sector Extension, Financial Agencies Extension, Market-led Extension, Contract Farming Extension, and NGO/CSO-led Extension. In the digital era, post-COVID extensions are shifting towards cyber extension methods such as Farmer Portals, Apps, DD Kissan, AIR, m-Kissan, etc., to disseminate important messages to a wider audience. Gender-sensitive extension models like the Women Para Extension Worker Model and Mission Shakti are also making an impact in Eastern India. Cooperative and FPO-based group-led extensions, along with agriculture fairs, exhibitions, awareness campaigns, diagnostic visits, surveillance, mass extension approaches, demonstrations, field days, training sessions, and meetings, are among the commonly used group approaches in Eastern India. These emerging approaches have redefined the role of extension services as a bridge between researchers and farmers, becoming a pivotal link among stakeholders for collaborative, participatory research-extension convergence strategies aimed at delivering efficient and effective extension services in Eastern India.

Introduction

Challenges in agriculture and natural resource management in the past have prompted changes in agricultural education, research, and extension globally. The sustainable development of agriculture in a country largely depends on the effectiveness of its agricultural extension strategies, encompassing approaches, service delivery, methodology, and processes. The desired outcome of agricultural development is farmers' outcomes, achievable through engagement, learning, and participation in research and extension processes. Agricultural extension played a significant role in the green revolution by providing farming communities with appropriate technologies and optimal cropping practices. This helped countries transition from being food deficit to food exporting (Ratnakar, 2016). Additionally, extension services play a crucial role in expanding knowledge, skills, and attitudes about the management of socio-economic-agricultural practices by understanding the existing techno-political ecosystem and exploring a basket of technology alternatives for making informed decisions.

The task force on the 'Hunger of the Millennium Project' recommended increasing agricultural productivity among food-insecure farmers by enhancing soil health, expanding small-scale water management systems, improving the accessibility of quality seeds, diversifying farm enterprises, and establishing an effective extension service (Ratnakar, 2016). Over the past five decades, extension services have undergone significant evolution, transitioning from their initial purpose of educating farmers about new agricultural technologies. Today, they are recognized as vital drivers of rural innovation and development. Emerging approaches have reshaped extension's role as a conduit between researchers and farmers, expanding beyond its original concept rooted in the one-way transfer of technology from developers to users during the 1960s. Valuable insights from various fields, including participatory research and extension, adult education, rural empowerment, farming systems research and extension, agricultural knowledge and information systems (AKIS), and the more recent agricultural innovation systems (AIS), have influenced and enriched the practice of extension (GFRAS, 2012).

Extension approaches encompass the broader strategies or frameworks utilized in the design, implementation, and evaluation of agricultural extension programs. These approaches are conceptual in nature, focusing on the underlying principles and values guiding the delivery of extension services. Common extension approaches include participatory approaches, farmer-led approaches, and value chain approaches. On the other hand, extension tools consist of specific techniques, methods, and resources employed by extension workers to transfer knowledge and information to farmers and other stakeholders. These tools may include demonstration plots, farmer field schools, mobile extension units, audio-visual aids, and mobile apps, among others. Extension tools are designed to aid extension workers in communication with farmers and facilitate the adoption of new agricultural technologies and practices. In summary, extension tools serve as the specific instruments used by extension workers to deliver information and knowledge to farmers, while extension approaches serve as the overarching strategies employed to design and deliver extension programs. Both tools and approaches are vital components of agricultural extension programs and should be carefully selected and tailored to meet the needs of specific farming systems and communities.

Currently, in eastern India, there is no exclusive approach, such as public or private. Instead, a multi-level partnership, convergence, franchising, community involvement, and ICT-mediated pluralistic extension system exist. This classification of extension approaches aims to provide a broader understanding of current extension dynamics, taking into account the heterogeneity and diverse demands of different farming systems.

1. Frontline Extension

Frontline extension is a research institute-led extension approach supervised closely by scientists. It was established by the Indian Council of Agricultural

Research (ICAR), which formed the Division of Agricultural Extension at its headquarters in 1971. Through its extensive network of research institutes, Agricultural Technology Application Research Institutes (ATARIs), Krishi Vigyan Kendras (KVKs), and Universities, ICAR disseminates modern farm technologies and advisory services via the first-line extension system and various outreach programs. ICAR's first-line extension services were initiated through programs such as the All India Coordinated Research Project on National Demonstration (1964), Operational Research Project (1974), Krishi Vigyan Kendras (1974), and Lab to Land Program (1979), among others. These programs aimed to bridge the gap between research findings and practical application in the field. Recently, ICAR has continued its first-line outreach extension programs under different project names, maintaining its commitment to disseminating agricultural innovations and knowledge to farmers across the country.

1.1 ICAR Outreach Extension

1.1.1 Farmer FIRST

Farmers often encounter challenges related to production and natural resource management for which they may not have readily available solutions. In such circumstances, the Farmer FIRST Programme (FFP) provides an opportunity for researchers, extension personnel, and farmers to collaborate and identify appropriate strategies through the assessment of various solutions. The Farmer FIRST initiative aims to empower farmers by equipping them with the necessary knowledge, resources, and support to enhance their agricultural practices and livelihoods. The acronym "FIRST" represents farmers' "Farm, Innovations, Resources, Science, and Technology," highlighting its focus on enriching the interface between farmers and scientists, assembling and applying technology, fostering partnerships, building institutions, and mobilizing content. This initiative was launched in October 2016 in India to address the challenges faced by farmers, particularly small and marginal ones. Farmer FIRST provides a platform for farmers and scientists to establish linkages, develop capacities, adapt and apply technology, manage on-site inputs, provide feedback, and build institutions. In the year 2016-17, fifty-two projects were sanctioned with a total outlay of Rs. 1653.60 lakh, benefiting approximately 45,000 farmers (ICAR-FFP, 2017). It is essential for these programs to be adapted to the specific conditions of each farming system and to involve active participation from both farmers and scientists.

1.1.2 Mera Gaon Mera Gaurav

The Mera Gaon Mera Gaurav-MGMG (My Village My Pride) scheme was launched in 2015 by ICAR as an innovative initiative to facilitate direct interaction between scientists and farmers, thereby expediting the lab-to-land process. The primary objective of this scheme is to furnish farmers with essential information, knowledge, and advisories on a regular basis by assigning groups of scientists to adopt villages. These groups are formed at both the institute and university levels.

Typically, a group of four scientists is designated at each institute/university to adopt villages situated within a radius of 50-100 kilometers from their workplace. These scientists collaborate with KVKs, Panchayats, and other relevant departments to secure necessary cooperation. Subsequently, they analyze the farming practices, climate conditions, as well as social and economic aspects of the selected villages. Based on their assessments, scientists formulate and recommend strategies tailored to the specific needs and circumstances of each village (ICAR-ATARI, Hyderabad, 2023).

1.1.3 NARI Nutri Sensitive Agriculture

The Indian Council of Agricultural Research has taken proactive steps to address women's nutrition concerns through a well-structured initiative called the 'Nutri-sensitive Agricultural Resources and Innovation (NARI)' program. This program adopts a food-centric approach, emphasizing the importance of nutrition-rich food crops, dietary diversity, and food fortification in agricultural development. The primary objectives of the NARI program, as outlined by the National Academy of Agricultural Sciences (NAAS, 2022), are as follows:

1. Encouraging Nutri-sensitive agriculture and raising awareness among female farmers and youth about Nutri-sensitive agricultural practices, including the cultivation of Nutri-gardens and promotion of Nutri-thalis.
2. Providing training and demonstrations to Anganwadi workers, who are predominantly women, on cultivating biofortified crop varieties, millets, and preparing nutritious recipes for Nutri-thalis.
3. Establishing location-specific Nutri-gardens and Nutri-smart villages aimed at promoting healthy and diverse diets accessible to all, with the assistance of KVKs, state Departments of Women and Child, and other relevant stakeholders.

1.1.4 VATICA

NITI Aayog (National Institution for Transforming India) reports that India faces nearly Rs 90,000 crore in annual post-harvest losses. Despite being the world's second-largest producer of fruits and vegetables after China, only 2% of this produce is processed. The processing rate remains low, below 10%, even though there is significant production capacity. Specifically, only about 2% of fruits and vegetables, 8% of marine products, 35% of milk, and 6%

of poultry undergo processing. ICAR has launched the "Value Addition and Technology Incubation Centers in Agriculture (VATICA)" initiative to tackle this issue. VATICA aims to spread awareness about post-harvest technologies and improve the skills of farmers, farm women, rural youth, and farmer organizations in different post-harvest management techniques. Additionally, VATICA will provide guidance and technical support to farmers and young individuals looking to start their post-harvest processing businesses (Kumar and Nangyal, 2020).

1.1.5 KSHAMTA

The tribal regions in our country possess a unique and distinct culture due to their geographical isolation. However, these areas still lack development, with agriculture serving as the main source of sustenance. To address this issue, ICAR has launched a specialized integrated program called "Knowledge Systems and Homestead Agriculture Management in Tribal Areas (KSHAMTA)" in 125 districts nationwide where the tribal population comprises at least twenty-five (25%) per cent.

The fundamental principle of KSHAMTA is to cultivate food that aligns with the preferences of the local people. This initiative focuses on leveraging the traditional knowledge of indigenous communities, such as preserving valuable medicinal trees, to promote agricultural development in the region. The primary goals of these programs include organizing activities related to nutritionally enriched foods and providing scientific support in areas like livestock and fisheries. Additionally, KSHAMTA involves mapping the entire food system in villages and offering recommendations on suitable dietary choices (Kumar and Nangyal, 2020).

1.2 SAU Directorate of Extension Outreach Programme

The Directorate of Extension Education (DOEE) serves as the central coordinating agency for State Agricultural Universities (SAUs), facilitating agricultural development within the state. This is achieved through the efficient transfer of technology, which includes providing training, consultation services, and disseminating farm-related information to both professional extension personnel within line departments and farmers. Additionally, the DOEE actively participates in evaluating, enhancing, and implementing technology through on-farm testing and demonstrations in the field. The directorate plays a crucial role in guiding, overseeing, and assessing the extension programs conducted by KVKs affiliated with SAUs. Furthermore, the DOEE supports state departments by distributing agricultural knowledge through publications covering various agricultural disciplines and related subjects. Therefore, the primary operational domains of the DOEE revolve around training, consultation, and communication (Singh *et al.*, 2013).

The DOEE consists of a team of experts from diverse fields who collaborate in a participatory manner and maintain close coordination with entities such as the Department of Agriculture, Animal Husbandry, Horticulture, Forestry, Cooperatives, Panchayat Samities, and other organizations committed to improving the lives of rural communities.

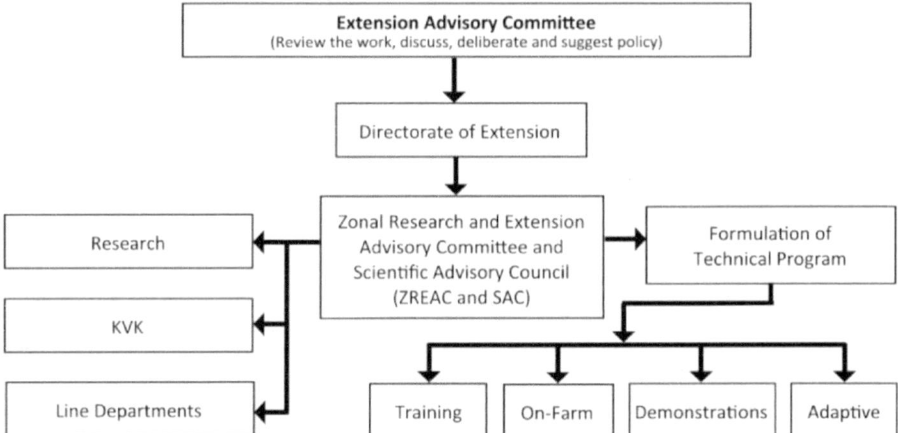

Fig. 1: Approaches and Methods used by the Directorate of Extension Education (Adapted from Singh *et al.*, 2013)

1.3 Krishi Vigyan Kendra (KVK)

The role of Krishi Vigyan Kendras (KVKs) has evolved significantly, shifting from providing vocational training to focusing on Technology Assessment and Demonstration for Application and Capacity Development in a district, thereby making a substantial contribution to India's National Agricultural Research and Extension System (NARES). While technology generation primarily takes place at ICAR research centers or State Agricultural Universities, KVKs play a crucial role in evaluating technology for location-specific suitability, refining it based on on-farm real-world challenges, and validating it for widespread adoption by farmers.

The Technologies Assessment and Refinement (TAR) process involve on-farm trials (OFTs) with active participation from farmers and a comprehensive on-farm analysis by KVK scientists. Following TAR, compatible and effective technologies are integrated into the extension system for broader adoption through Front Line Demonstrations (FLDs) closely supervised by KVK scientists and extension officials. KVKs also conduct capacity development programs for farmers, farm women, rural youth, NGOs, and extension personnel, forging links with allied agencies to offer complementary services. Moreover, KVKs serve as a single window delivery system and Knowledge Resource Center (KRC), leveraging ICT methods to disseminate technological products and advice to a wider audience within the district.

Under the National Food Security Mission (NFSM), in collaboration with ICAR, KVKs are involved in the Seed Hub project, empowering farmers in indigenous pulse and oilseed seed production with technical support. Additionally, through Cluster Front Line Demonstrations (CFLDs), KVKs encourage farmers to expand pulse and oilseed cultivation, particularly in rainfed areas, enhancing crop diversity. KVK scientists lead and supervise these demonstrations. Furthermore, 199 KVKs operate as District Agro-met Units (DAMU), providing agro-met advisory services in local languages across different agro-climatic regions for climate-resilient agriculture and sustainable production.

The ARYA (Attracting and Retaining Youth towards Agriculture) initiative, initiated by ICAR, empowers rural youth in agricultural and allied skill development through KVKs, creating income-generating opportunities in the agricultural sector. These trained youths can serve as role models and master trainers for other rural youths interested in agricultural start-ups. KVKs collaborate with ICAR institutes to provide technology support and initial guidance for incubating trained youth in agriculture-related enterprises. Hence, KVKs bridge the gap between research and extension, acting as a convergence hub for all stakeholders working within the district to provide optimal services to the farming community.

1.4 ATIC of ICAR and SAU

The Agricultural Technology Information Centre (ATIC) functions as a comprehensive support system, serving as a central point of contact that bridges the gap between various research institutions, intermediary users, and end users (farmers) in decision-making and problem-solving efforts. Numerous valuable agricultural technologies, including knowledge, technologies, seeds, planting materials, and publications, have been developed in various institutions. However, these resources can fulfil their purpose effectively only when they are put into practice by the right individuals, and ATIC plays a crucial role in meeting these needs. ATIC is staffed by multidisciplinary experts from SAUs or research institutes who cater to farmers' needs. It serves as the face of the institute, conducting live demonstrations of technologies in the cafeteria, selling inputs, diagnosing disease and pests for farmers, providing advisory services, and gathering feedback from end-users for researchers to enable continuous improvement. Additionally, ATIC acts as a repository of agricultural information related to farming practices, skills, farm inputs, and agricultural education through publications and digitization efforts (Anonymous, 2023).

1.5 Agri- Business Incubation Centers (ABIC) of ICAR and SAU

The ABI Centre serves as a platform where individuals can connect with experts in their respective fields to receive guidance on entrepreneurship development and strategies to make farming more profitable. This center offers two modes of incubation:

1. On-Campus Incubation: Aspiring entrepreneurs are provided with incubation space and comprehensive support related to their commodity of focus, such as rice in the crop-based ecosystem of ICAR-National Rice Research Institute (NRRI), Cuttack.
2. Off-Campus Incubation: In cases where on-campus space is limited, the center extends support through off-campus incubation. This mode is available for Farmer Producer Companies (FPCs) and other agricultural startups not centered on the specific commodity of interest. For instance, support for non-rice crop cultivation at NRRI-ABI.

The Agribusiness Incubation Centre conducts various skill-based agri-preneurship training programs aimed at developing the potential of farmers, entrepreneurs, ex-servicemen, educated and unemployed rural and urban youths as agri-preneurs (ICAR-NRRI, 2023). These training programs focus on imparting managerial and entrepreneurial skills to aspiring candidates. Different programs are available based on requirements:

a. Comprehensive Agribusiness Incubation Program (3 weeks)
b. Technology-based Entrepreneurship Development Program (1 week)
c. Skill-based Entrepreneurship Development Program (1 week)
d. Training on Demand (1 week)

The Vyapar Initiative in Agriculture and Agri-Startups, known as VIKAS-RABI and supported by RKVY-RAFTAAR, is established within agricultural institutes, including ICAR-NRRI, Cuttack in Odisha. VIKAS-RABI aims to foster entrepreneurship and the creation of agricultural and allied sector businesses through innovation. It focuses on skill development, capacity building, and technology scaling through collaborations with industry and government partners. The program provides training by commercialization experts in the agricultural sector, access to a mentor network, and guidance from technical experts to help incubate innovative ideas and scale them up from ideation/start-ups to a larger scale (Vikasrabi.com, 2023).

1. Government Ministry based Extension

2.1 National Mission on Agriculture Extension Technology (NMAET): The National Mission on Agriculture Extension Technology (NMAET) comprises four sub-missions aimed at enhancing agricultural extension, seed and planting material supply, agricultural mechanization, and plant protection and quarantine.

2.1.1 Sub-Mission on Agricultural Extension (SAME): This sub-mission focuses on increasing awareness and promoting the adoption of suitable agricultural technologies in the agriculture and related sectors. It aims to build on past achievements by expanding the outreach of extension professionals. Trained individuals from programs like Agri-Clinics and Agri-Business Centres Scheme (ACABC) and Diploma in Agriculture Extension Services for Input Dealers (DAESI) will assist in delivering these services to farmers.

2.1.2 Sub-Mission on Seed and Planting Material (SMSP): This sub-mission focuses on ensuring the availability of high-quality seeds, which is crucial for boosting agricultural production and productivity. It covers the entire seed supply chain, from nucleus seeds to certified seeds provided to farmers, along with activities related to seed resource development. It also strengthens the Protection of Plant Variety and Farmer Rights Authority (PPV & FRA) to protect plant varieties and farmer rights and promote the development of new plant varieties through breeding.

2.1.3 Sub-Mission on Agricultural Mechanization (SMAM): Farm power availability significantly impacts farm productivity. SMAM is dedicated to agricultural mechanization, including custom hiring, mechanization of selected villages, and grants for tools and equipment acquisition to meet the needs of small and marginal farmers.

2.1.4 Sub-Mission on Plant Protection and Plant Quarantine (SMPP): This sub-mission aims to increase agricultural production by ensuring pest and disease-free crops through scientifically sound and environmentally friendly Integrated Pest Management (IPM) techniques. It involves enhancing pest management facilities, quality control of agricultural protection chemicals, disease-free plant production, and monitoring pesticide residues in food products. Bio-security measures for sustainable plant health management and capacity-building efforts are also part of this sub-mission.

Skill training for farmers and regional extension activities under these sub-missions will align with similar programs implemented by the Agriculture Technology Management Authority (ATMA).

Illustrative list of Farmer Centric Traninings and Field Extension			
SMSP	**AMAE**	**SMAM**	**SMPP**
Seed Village Programme	Farm Schools, Demo Plots, Trainings, Exposure Visits	Capacity Building by Institutions Identified by the State Government	Pest Monitoring (including Pest Scouts) FFSs, IPM Training to Farmers

Fig. 2: Extension tools adopted and convergence in NMAET submissions (EEI, Jorhat, 2019)

2.2 State Department Extension

In Indian states, agricultural extension, which may be considered conventional or ministerial extension, falls under the purview of the state government since agriculture is a state subject. Although the national government has an agriculture ministry and department overseeing broader national extension programs across all states and union territories, the primary responsibility for extension services lies with the state government. Below are some examples of extension programs in the state of Odisha as cited in official documents (FARD, GoO, 2018; DAFE, GoO, 2022).

2.2.1 Agricultural & Horticultural Extension

In eastern Indian states like Odisha, the agriculture ministry operates through the agriculture department to serve the farmers of the state. Several state-owned schemes and programs are implemented, including:

- Krushak Assistance for Livelihood and Income Augmentation (KALIA) Scheme
- Bhoomihina Agriculturist Loan and Resources Augmentation Model (BALARAM) Scheme
- Mukhyamantri Krushi Udyog Yojana (MKUY)
- Crop Diversification Programme (CDP)
- Capital Investment and Farm Mechanisation
- Crop Production Management
- Farmer's Welfare and Innovative Projects
- Harnessing Surface and Ground Water
- Information, Education, and Communication
- Training and Capacity Building in the agriculture sector of Odisha

In the horticulture sector, various schemes are also in place, including:

- Input Subsidy support
- Implementation of Horticulture Programme in Non-Horticulture Mission District
- Development of Potato, Spices & Vegetables
- State Potato Mission
- Promotion of Agriculture Production Clusters (APCs) in tribal regions of Odisha

2.2.2 Animal Husbandry & Livestock Extension

The Veterinary and Animal Husbandry Department plays a crucial role through its major arms: the Diagnostic & Treatment Wing and the Livestock Production & Management Wing, which directly contribute to the animal husbandry extension system. Additionally, the fishery sector focuses on aquaculture extension for fishery farmers. Various programs are implemented in several areas, including:

- Dairy Development
- Small Animal Development
- Poultry Development
- Fodder Development
- Livestock Health Care Service Delivery
- Training & Capacity Building
- Animal Welfare (State SPCA)
- Odisha State Poultry Products Co-operative Marketing Federation Ltd. (OPOLFED)
- Odisha State Cooperative Milk Producers Federation Ltd. (OMFED)

In Odisha, there are 541 veterinary hospitals/dispensaries and 3239 livestock aid centers providing veterinary services. The Central Clinics of the College of Veterinary Science and Animal Husbandry in Bhubaneswar offer outdoor treatment, doorstep health services, and artificial insemination services for a fee. They also provide indoor facilities for small animals and modern diagnostic services.

Various schemes in operation under the Fisheries sector in the state of Odisha are:

- Input Assistant to WSHGs for Pisciculture in Gram Panchayat tanks
- Machha Chasa Pain Nua Pokhari Khola Yojana

- Promotion of Intensive Aquaculture through Bio-floc Technology
- Livelihood Support to Marine Fisherman during Fishing Ban Period
- New Tank Excavation under Brackish Water Aquaculture
- Assistant to Network of Seed Growers under Early Breeding Programme

2.2.3 Soil Conservation and Watershed Extension

The Soil Conservation and Watershed Extension department in Odisha is instrumental in implementing various schemes aimed at sustainable agricultural practices and environmental conservation. This includes Farm Pond, Farm Pond+, Innovative Agroforestry for Food and Nutrition Security, Rejuvenating Watersheds for Agricultural Resilience through Innovative Development (REWARD), Odisha Mineral Bearing Area Development Corporation (OMBADC), District Mineral Foundation (DMF), FAO-GEF assisted Green Agriculture Project, etc.

3. Capacity Building Extension

3.1 National Institute of Agricultural Extension Management (MANAGE)

MANAGE, an autonomous organization operating under the Ministry of Agriculture & Farmers' Welfare, Government of India (GoI), was established to address the evolving challenges within the rapidly expanding agriculture sector. Its primary mission is to support the Government of India and State Governments/Union Territories in enhancing the efficiency of agricultural and allied sectors by tailoring policies and programs to address specific needs. Additionally, MANAGE seeks to bolster the capabilities, expertise, and mindset of extension personnel. The institute's core areas of focus encompass training, education, research, consultancy, documentation, and the implementation of select Central Sector Schemes.

MANAGE, being an apex level institute in agricultural extension management, facilitates various activities under the Support to State Extension Programmes for Extension Reforms (ATMA) Scheme (DoE, DAFW, MoA, GoI, 2018). MANAGE actively identifies brief training programs lasting 7-10 days to motivate scientists and extension officers engaged at the state, central, and union territory levels,. In collaboration with national and international agricultural extension institutes, MANAGE obtains their annual training schedules and shares them with state governments to solicit nominations. The received nominations undergo a thorough review to select qualified extension personnel for studying exemplary extension methodologies and participating in training programs offered by prestigious institutions in India and overseas.

MANAGE has introduced an innovative initiative known as the "Certified Farm Advisor/Certified Livestock Advisor" program, aimed at transforming agricultural extension personnel into specialists in specific crop or livestock areas. This program comprises three modules covering fundamental aspects of the latest technologies in crop and livestock management. Candidates participating in the programme are required to complete a three-month online training course facilitated by MANAGE. After completing rigorous training, candidates are allotted Mentor Scientists from the concerned Research Institute for one year for technical guidance to face field-level challenges. On completion, candidates' knowledge is assessed based on identified parameters to be declared as "Certified Farm Advisor" or "Certified Livestock Advisor" by MANAGE and its technical partner. Post Graduate Diploma in Agricultural Extension for eligible students or in-service extension functionaries continues to be implemented through MANAGE, receiving central share directly (MANAGE, 2023).

3.2 National Institute of Plant Health Management (NIPHM)

The National Institute of Plant Health Management (NIPHM) is dedicated to enhancing the effectiveness of pest and disease monitoring and control systems, as well as certification and accreditation procedures, for Indian states and the national government. NIPHM serves as a central hub for training, adaptive research, extension services, and policy development related to plant protection, catering to both public and private sector organizations.

In addition to conventional training initiatives, NIPHM offers Post Graduate Diploma in Plant Health Management courses, conducts projects, engages in capacity-building activities, and conducts research in the field of plant health and quarantine. This includes evaluating market access potential and addressing various aspects related to the Sanitary and Phytosanitary Measures (SPS Agreement). NIPHM also sponsors and regulates a 12-day Certificate Course on insecticide management for insecticide dealers/distributors, conducted at every KVK or District Training Centre (DTC), making them nodal training centers for districts (NIPHM, DAFW, MOA, GOI, 2023).

3.3 Indian Institute of Plantation Management (IIPM)

The Indian Institute of Plantation Management (IIPM) in Bengaluru was established in November 1993 under the Ministry of Commerce & Industry (MoC&I), Government of India, as an independent institution of higher education. Initially, the institute worked closely with the plantation industry and the Commodity Boards of India (CBI) to support the development of the agricultural plantation sector. IIPM offers programs such as Agribusiness

Management (ABM), Post Graduate Diploma courses, short-duration capacity building, and training programs designed to meet the specific needs of stakeholders in the agri-plantation sector. These programs cater to a diverse audience, including plantation owners, corporate managers, executive members of planter associations, small and medium-sized entrepreneurs, officials, and scientists from the CBI, various Ministries of the Government of India, as well as officials from State Development Departments (IIPM, 2023).

3.4 Extension Education Institute (EEI)

The Extension Education Institute (EEI) operates as a regional-level institution with the primary objective of providing capacity building to middle-level extension functionaries in the field of Extension Education within departments and organizations of agricultural development. The Department of Agriculture and Cooperation (DAC), Ministry of Agriculture, Government of India, has established four EEIs across different regions of the country: EEI (Southern Region) in Hyderabad, Telangana; EEI (Northern Region) in Nilokheri, Haryana; EEI (Western Region) in Anand, Gujarat; and EEI (NE Region) in Jorhat, Assam.

The objectives of these institutes, as outlined by Jana (2016), include:

- Providing in-service training to staff from State Training Institutes, Line Departments, and State Agricultural Universities in Extension Teaching Methods and Communication Media.
- Organizing workshops on Communication and Extension Teaching Methods/Training Methodology for Master Trainers, Sub Divisional Agricultural Officers, and Subject Matter Specialists involved in Broad-based Agricultural Extension.
- Conducting workshops in specialized fields such as Monitoring and Evaluation, Supervision, and Extension Management for Middle Level Extension personnel engaged in Broad-based Agricultural Extension.
- Undertaking programs for the publication and production of basic teaching/training material relevant to extension personnel.
- Conducting continuous field studies on Extension Education and related subjects.

3.5 State Agricultural Management Extension Training Institute (SAMETI)

SAMETI operates as a state-level autonomous institution focused on capacity building in technical and human resource management within agriculture. It arranges customized training programs for individuals involved in project implementation across various government departments and the farming community. SAMETI functions under the technical guidance of the National Institute of Agricultural Extension Management and operates in each state under different names like BAMETI for Bihar and IMAGE for Odisha.

As an autonomous entity, SAMETI performs several key functions (SAMETI, 2023):

1. Provides consultancy services tailored to the specific needs of the Agricultural Technology Management Agency (ATMA) in project planning, appraisal, and implementation.
2. Develops and promotes specialized management tools to enhance the efficiency of agricultural extension services by managing human and material resources effectively.
3. Organizes customized training programs for middle-level and grassroots agricultural and allied extension professionals, focusing on management, communication, participatory methodologies, and more, based on feedback from previous training initiatives.
4. Establishes close partnerships with institutions such as KVKs, Zonal Research Stations (ZRS), State Agricultural Universities, Non-Governmental Organizations (NGOs), and management institutions like MANAGE. These partnerships enable SAMETI to access appropriate faculty resources from these institutions for training and consultancy services to ATMA personnel, farmers, and other stakeholders.
5. Conducts research studies on issues related to Agricultural Extension Management, Communication, Information Technology, Agricultural Product Marketing, and Human Resource Development using participatory approaches.
6. Emerging Public Sector Participatory Extension

4.1 Agricultural Technology Management Agency (ATMA)

The Centrally Sponsored Scheme known as the 'Support to State Extension Programs for Extension Reforms' or the ATMA Scheme has been operational since 2005 and is currently active in 691 districts across 28 states and five Union Territories in India. This scheme aims to establish a decentralized, farmer-friendly extension system nationwide. Its main objectives include supporting state governments in providing farmers with the latest agricultural technologies and best practices through various extension activities such as farmer training, demonstrations, exposure visits, Kisan Mela, mobilizing farmers' groups, and organizing farm schools. The Agricultural Technology Management Agency (ATMA) is a district-level registered society that collaborates with various agricultural institutions to promote sustainable agriculture development. Its primary objectives include coordinating research and extension activities at the district level and decentralizing the public agriculture technology system. One key goal is to establish "Farmer's Interest Groups (FIGs)" with the involvement of public organizations, private organizations, NGOs, Para Extension Workers, and private input dealers in all the blocks and villages of the district.

The objectives of the ATMA Scheme are as follows (My Schemes, GoI, 2023):

- Implementing a bottom-up approach with decentralized decision-making based on the Annual Action Plan of blocks, district, and state plans along with the 5-year Strategic Research Extension Plan (SREP).
- Establishing Block ATMA Cells consisting of Block Technology Teams (BTT) and Block Farmers' Advisory Committees (BFAC) to prepare the Block Action Plan (BAP).
- Emphasizing a group approach to extension, including Farmers' Organizations (FOs), Farmer's Interest Groups (FIGs), Commodity Interest Groups (CIGs), Self-Help Groups (SHGs), etc.
- Addressing gender concerns with a minimum 30% allocation in all expenses.
- Converging line departments in planning, scheme development, and execution.
- Implementing multi-agency extension strategies involving NGOs, Farmer Producer Organizations (FPOs), Krishi Vigyan Kendras (KVKs), State Agricultural Universities (SAUs), and other allied departments.
- Appointing local progressive farmers as Farmers' Friends with minimal remuneration.
- Adopting the Farm School approach on progressive farmers' fields.
- Recruiting agricultural graduates as Block Technology Managers and Agricultural Technology Managers to facilitate all extension programs in the block.
- Ensuring the District Magistrate serves as the Chairman of ATMA to review action plans, fund expenses, and mobilize multi-agencies working in ATMA.

Extension methods used in the ATMA Scheme include training in agriculture and related subjects, demonstrations, exposure visits inside and outside the state, Krishi Melas/Exhibitions, Farmer-Scientist Interactions facilitated by KVK resource persons, farm schools on progressive farmers' fields, Best ATMA

Farmer's Awards, publication of agriculture-related materials, encouraging public-private partnerships, promoting women's participation in agriculture, publishing and sharing success stories in the agriculture and allied sectors, and organizing innovative folk shows like Kalajatha for seed treatment campaigns and package of practices demonstrations.

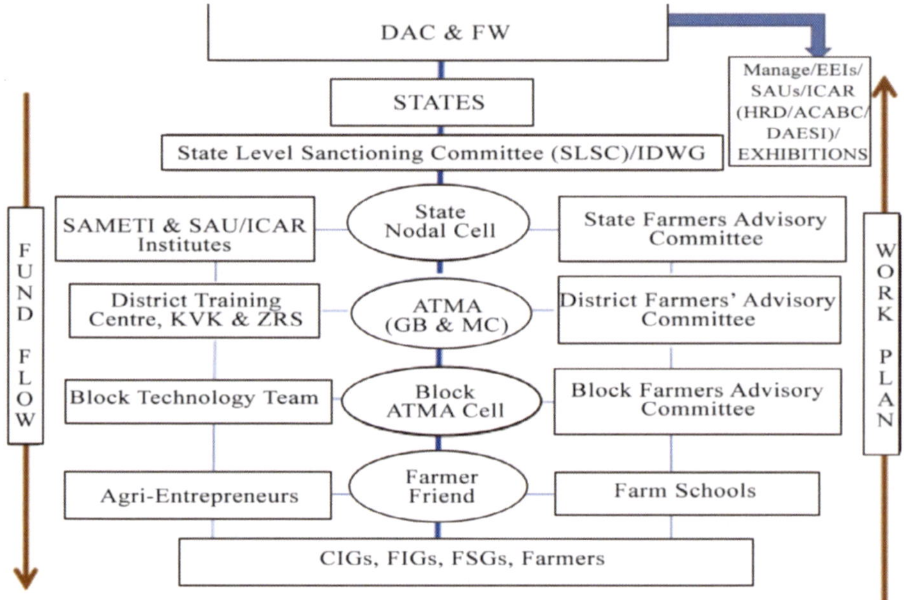

Fig. 3: ATMA structure (*Source*: DoE, DAFW, MOA, GoI, 2018)

2. Franchised Convergence Extension

5.1 Franchising out to other Public Sectors

Franchising out government agricultural extension services in Eastern India involves forming partnerships with other public sector organizations to outsource certain extension activities. This collaboration allows for the utilization of the expertise, resources, and wider reach of these entities to enhance the delivery of agricultural extension services. The key benefits of this partnership include collaboration, complementary services, access to expertise, technologies, and training, all of which contribute to increasing the efficiency and effectiveness of agricultural extension efforts.

5.1.1 IARI-Post Office Extension Linkage Extension Model

The ICAR-Indian Agricultural Research Institute (IARI) introduced the IARI-Post Office Linkage Extension Model in 2009 to extend the public extension system to farmers across India's diverse climatic and geographic regions. This model, initially piloted in Sitapur district, Uttar Pradesh, was later expanded to 55 districts in 14 states, including eastern India, enhancing the national extension network.

Under this model, village postmasters and farmers receive training in improved farming practices through collaboration with local KVKs. Demonstrations of improved IARI varieties on the farms of village postmasters serve as learning platforms for other farmers, encouraging "Farmer to Farmer" seed sharing. Postmasters act as para-extension agents for ICAR-IARI with KVKs' support. Research by Sahoo *et al.* (2017) identified "reaching distant farmers" as a significant strength, along with "timely delivery of seeds" and "cost-effectiveness."

ICAR's location-specific extension model leverages its research institutes' proximity to crop locations, reducing transport and postal transaction costs for farmers. SAUs, state departments, and corporations could adopt similar approaches, utilizing rural postmasters as extension agents in collaboration with the postal department.

5.2 Government-linked Agri-business Corporation

5.2.1 Agricultural Promotion and Investment Corporation of Odisha Limited (APICOL)

The Agricultural Promotion and Investment Corporation of Odisha Limited (APICOL), a government undertaking in Odisha, plays a crucial role in promoting and supporting agricultural enterprises in the state. It provides guidance and assistance for project formulation, counseling, enterprise development, and project implementation support. APICOL is actively involved in implementing various schemes such as the Mukhyamantri Krushi Udyoga Yojana (MKUY), Agriculture Entrepreneurship Promotion Scheme (AEPS), Prime Minister Formalization of Micro Food Processing Enterprises (PMFME), and Agriculture Export Policy (AEP) on behalf of the government.

Over the years, APICOL has provided significant financial support to 1972 Commercial Agri Enterprises (CAEs) and 4587 Agro Service Centres (ASCs) in Odisha, amounting to approximately Rs. 308 crores in subsidies. Moreover, it has facilitated irrigation projects under the Jalanidhi scheme by distributing subsidies worth Rs. 650.30 crore for 2,28,982 shallow tube wells, dug wells, bore wells, and lift irrigation projects. This initiative has contributed to the irrigation of 5.725 lakh hectares of agricultural land in the state.

APICOL employs various extension methods such as entrepreneur counseling, training sessions, seminars, exhibitions, and project report preparation to disseminate information and support agricultural development in Odisha.

5.2.2 Odisha Agro Industries Corporation (OAIC)

OAIC operates with the objective of marketing agricultural inputs and farm machinery through an extensive network of offices at district and sub-district levels. The corporation plays a crucial role in executing Tube Wells, Bore Wells, and Lift Irrigation Points, as well as supplying diesel pump sets, sprinklers, drips, pipes, crates, polyethylene, and other essentials to government organizations for farming communities. Notably, OAIC contributes significantly to schemes such as Jalanidhi, irrigation projects, farm mechanization, and horticultural programs, among others (Odishaagro.nic.in)

5.2.3 Odisha State Seeds Corporation (OSSC)

The Odisha State Seeds Corporation (OSSC) serves as the designated Nodal Agency responsible for producing, procuring, and distributing certified quality seeds to farmers. It was established on 24 February, 1978, with the primary goal of manufacturing certified seeds and making them available to farmers at reasonable prices through Dealers and PACS/LAMPS using Direct Benefit Transfer (DBT) since the kharif-2016 season. To achieve this objective, the corporation conducts seed production programs with seed growers across various locations in Odisha under the "Mo Bihana Yojana" scheme (OSSC Ltd., 2023).

5.3 Franchising out to International Research Institute (CGIAR): (Dhar, 2016)

On 30 June, 2016, a partnership agreement was signed between the International Maize and Wheat Improvement Center (CIMMYT) and the Indian Council of Agricultural Research (ICAR), Government of India. This collaboration, known as the Cereal Systems Initiative for South Asia (CSISA) Phase III, aims to improve the productivity and income of smallholder farmers in eastern India. The focus is on sustainable intensification and enhancing extension systems. The partnership underscores the importance of updating traditional farming practices in India through participatory research involving KVKs. The ultimate goal is to develop location-specific extension strategies and investment plans integrated into the State Departments of Agriculture in Bihar, Uttar Pradesh, and Odisha. CSISA has played a pivotal role in establishing a collaborative "convergence platform" in Bihar, Odisha, and eastern Uttar Pradesh. This platform brings together various stakeholders, including state agro-development bodies, government departments, cooperatives, and private organizations, to streamline efforts, reduce duplication, facilitate swift decision-making, and promote mutual learning among stakeholders. Additionally, CSISA and ICAR Agri-Extension have helped set up state and district-level convergence groups. CSISA's initiatives in India cover a wide range of areas, including promoting directly-sown rice (DSR) to address labour and energy challenges, strengthening agro-advisory services through knowledge management and data integration, developing precision nutrient management strategies, encouraging income-generating maize production in unexplored areas, focusing on rice-fallow development in coastal Odisha, enhancing the capabilities of National Agricultural Research and Extension Systems (NARES), implementing integrated weed management practices, accelerating the adoption of mechanized solutions, and addressing challenges related to climate extremes in rice-wheat cropping systems.

1st image: KVK-CSISA Network Annual Workshop at NASC, New Delhi; **2nd image:** Exposure visit of farmers & extension officials of Odisha for mechanized DSR in Bapatla, Andhra Pradesh (Courtesy: csisa.org.in)

5.4 Collaborative Scheme Development and Execution

5.4.1 Bill & Melinda Gates Foundation and Government of Odisha: (Gates Foundation, 2023)

ADAPT (Analytics for Decision Making and Agricultural Policy Transformation), a collaborative effort involving the Foundation, the Odisha Government, and consulting firm Samagra started in 2017, aids in making data-informed agricultural policy decisions in Odisha. Its aim is to boost agricultural productivity by optimizing supply chains, providing pest and weather advisories, and disseminating market and technology information to farmers. Furthermore, ADAPT partners with the state government on the KALIA scheme, which supports smallholder and marginalized farmers. The ADAPT Dashboard consolidates various agricultural databases into a single online portal, enabling government officials to proactively address challenges and meet the needs of farmers. Farmers can utilize the Dashboard to access data on seasonal variations, soil health, and required inputs.

5.4.2 Bill & Melinda Gates Foundation and Government of Bihar

JSPVAT, the JEEViKA Special Purpose Vehicle for Agricultural Transformation, was launched in 2020 to bolster the Bihar state government's rural livelihoods initiative, JEEViKA. Its primary objective is to improve the income, nutrition, and sanitation of farming households in Bihar. JSPVAT has forged a significant partnership between JEEViKA and the World Bank to strengthen agricultural and livestock market ecosystems in the state. Collaborating with 50,000 farmers, JSPVAT focuses on implementing changes to support market connections, testing innovative private-sector models, and leveraging digital solutions to enhance procurement, quality testing, access to finance, and technology adoption. It also encourages the establishment of farmer producer companies to engage with smallholder farmers and has introduced initiatives to enhance market ecosystems for various agricultural products. Notably, JSPVAT introduced derivative trading on the NCDEX platform, enabling farmers to achieve prices 6 to 8 per cent higher than previously.

5.5 Community Participatory Extension

Seed Village Scheme

The "Seed Village Scheme", also known as the "Beej Gram Yojana", has been implemented by the Government of India since 2014-15 to enhance the quality of seeds saved by farmers, which make up around 80-85% of the total seeds used in crop production programs. Under this scheme, financial support is provided for the distribution of foundation/certified seeds, covering 50% of the seed cost for cereal crops and 60% for pulses, oilseeds, fodder, and green manure crops, up to one acre per farmer. Farmers are offered various resources such as training sessions, exhibitions, presentations, field demonstrations, farmer meetings, and similar events focused on seed production and technology dissemination (PIB, MOA, 2021). The seeds produced in these designated seed villages are preserved and stored until the subsequent planting season. To encourage farmers to build appropriate storage capacity, they receive assistance for constructing or acquiring storage bins such as Pusa Bins, Mud bins, or Bins made from paper pulp for storing seeds produced on their farms. The implementing agencies include State Departments of Agriculture, State Agriculture Universities, KVKs, State Seeds Corporation, National Seeds Corporation, State Farms Corporation of India (SFCI), State Seeds Certification Agencies, and the Department of Seed Certification. Farmers play a crucial role as active stakeholders in the successful execution of the program.

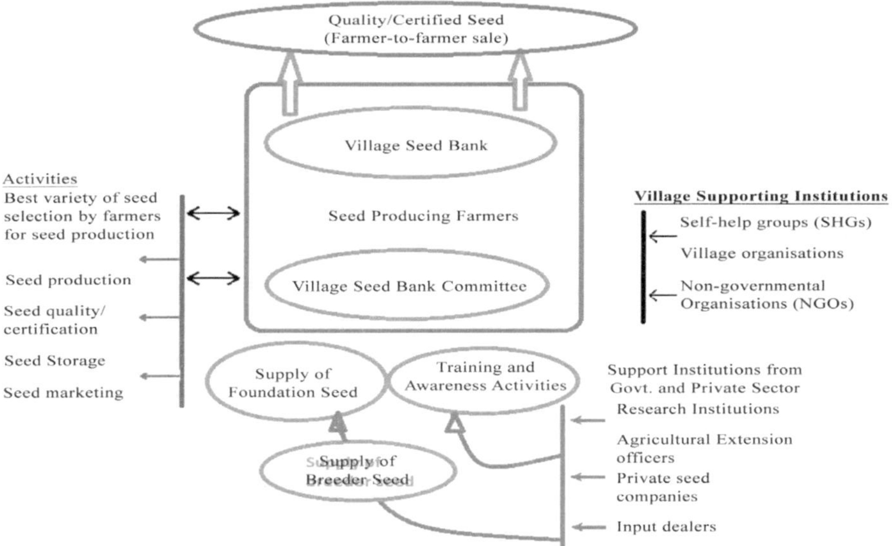

Fig. 4: Decentralized seed production under Seed Village Programme (SVP) (Adapted from Bhavani *et al.*, 2022)

6. Public Sector Target Approach Extension

6.1 Complex Diverse Rainfed Plan

The Rainfed Area Development (RAD) Scheme, initiated as part of the National Mission for Sustainable Agriculture (NMSA) during 2014-15, aims to promote sustainable development in rainfed areas. The scheme focuses on implementing Integrated Farming System (IFS) models developed by the Indian Council of Agricultural Research (ICAR) for different agro-climatic zones to enhance sustainable agricultural production. Key components of RAD include region-specific improved agronomic practices, soil health management, rainwater optimization, responsible chemical usage, crop diversification, and integrated crop-livestock-tree farming systems, all aimed at fostering sustainable agriculture in rainfed areas. To effectively serve the diverse agro-ecologies found in rainfed regions, there is a need to reorient and train extension functionaries. The proposed ratio of extension functionaries to farmers in rainfed regions is set at 1:1000, emphasizing a blend of human resources and information and communication technology (ICT). Training programs should focus on the latest advancements in rainfed agriculture technologies and Sustainable Agricultural Production Systems (SAPSs) to facilitate the adoption of resilient cropping systems and improve productivity. Grassroot-level personnel involved in watershed and rainfed agriculture development programs require specialized training to enhance their capabilities.

Model watersheds are instrumental in scaling successful watershed management approaches. These model watersheds, functioning as 'Pilot Replicable Watersheds' at the district level, demonstrate effective watershed management practices. Their development should involve collaboration among local institutions, including research, development, government bodies, and civil society, with a focus on scalability. Extension institutes should adopt model watersheds to facilitate scaling up interventions effectively. The institutional structure of the National Rainfed Area Authority (NRAA) includes five verticals, focusing on Water Management, Watershed Development, Agriculture/Horticulture, Animal Husbandry/Fisheries, and Forestry. However, to ensure comprehensive development in rainfed regions, additional verticals addressing Secondary Agriculture, Extension, and Farmers' Organizations could be introduced to enhance NRAA's service capabilities (NRAA, DAFW, MOA, GOI, 2022).

6.2 Tribal Sub Plan

The Tribal Sub Plan (TSP) is a government initiative aimed at allocating a portion of funds from the Union Budget towards the socio-economic development of tribal communities in India. Its main goal is to reduce development gaps and improve the living standards of tribal populations by promoting sustainable agricultural practices, technology adoption, and capacity building. Under the TSP Program at ICAR-NRRI, Cuttack, tribal farmers actively participate in 'On-farm Farmers Participatory Trials,' where various rice varieties and agricultural technologies developed by the institute and its regional stations are demonstrated. To empower stakeholders, the program conducts training sessions covering various aspects of crop management, including institute visits, farmer meetings (Kishan gosthis), and other events during the cropping season. These interactions between farmers and scientists address issues related to agricultural production and livestock rearing, leading to the development and adoption of suitable technologies for enhancing crop and livestock productivity sustainably. Additionally, the program has established a "Farm Inputs Users' Association" (FIUA), receiving agricultural implements like Seed treating drums, Manual-operated markers, Mandwa weeders, Pedal paddy threshers, and Mini rice parboiling units. These tools are utilized for demonstration and practical use by association members. Similarly, ICAR Research Institutes and several State Agricultural Universities (SAUs) receive TSP funds to support technological assistance and capacity building of tribal farmers in eastern Indian tribal-dominated states.

In 2021, KVKs under the Agricultural Technology Application Research Institute (ATARI) Kolkata achieved significant outcomes in the Tribal Sub

Plan component:

- They created 3343 different assets, such as sprayers, ridge makers, pump sets, weeders, storage bins, drip irrigation sets, and poultry feeders/ drinkers, benefiting tribal farmers.
- Conducted 96 On-Farm Trials (OFTs) and 1244 Frontline Demonstrations (FLDs), contributing to agricultural development by testing and showcasing practices and technologies.
- Provided training to over 12,000 farmers in modern agricultural practices.
- Organized various extension activities, engaging more than 45,000 farmers in workshops, field visits, and awareness campaigns.
- Produced significant quantities of agricultural inputs, including seed, planting material, and livestock strains, contributing to improved productivity and sustainability in the region.

6.3 Scheduled Caste Sub Plan (SCSP)

In India, Scheduled Castes (SC) farmers, positioned at the lower rungs of the social hierarchy, grapple with significant challenges arising from limited land ownership, which heightens their reliance on wage labor. Their situation is exacerbated by social discrimination and economic disparities, affecting their participation in both farm and non-farm labor markets. Moreover, traditional monoculture and agricultural practices fall short in meeting the increasing and diverse food and nutritional needs (ICAR-CRIJAF, 2022). ICAR-Central Research Institute for Jute and Allied Fibres (CRIJAF) in Kolkata has initiated the SCSP Program to uplift the socioeconomic status of SC farmers, with key objectives including promoting modern agricultural techniques, ensuring self-sufficiency with essential agricultural inputs, demonstrating proven technologies in Jute and allied fibres-based farming systems, and exploring value addition possibilities through rural micro-enterprises and SC Women Self-help Groups. Additionally, ICAR-Indian Institute of Water Management (IIWM) in Bhubaneswar organized training programs focusing on crop planning and management strategies for enhancing farm income in post-flood situations under the SCSP project (ICAR-IIWM, 2022). Similarly, ICAR-Central Institute for Women in Agriculture (CIWA) in Bhubaneswar conducted various capacity-building activities such as improved paddy cultivation, skill enhancement in vegetable cultivation, promotion of labor-saving farm tools, establishment of nutri-gardens, and beekeeping to double the income of farm families, particularly targeting SC women from Scheduled Caste families (ICAR-CIWA, 2020). Furthermore, ICAR-NRRI in Cuttack organized training programs on high-yielding rice varieties and improved farming practices for better livelihoods under the SCSP program (ICAR-NRRI, 2023), while ICAR-Central Inland Fisheries Research Institute (CIFRI) provided training on reservoir and wetland fisheries management to fishers in Odisha under the SCSP sub-plan.

6.4 State Plan (RKVY/RAFTAAR)

The Rashtriya Krishi Vikas Yojana (RKVY) scheme, launched in 2007, aimed to facilitate comprehensive development in agriculture and associated sectors. Over time, it has undergone evolution and was operational during the 11th and 12th Plan periods. The scheme encourages states to increase investments in agriculture and related domains. In 2017, it transformed into the Rashtriya Krishi Vikas Yojana-Remunerative Approaches for Agriculture and Allied Sector Rejuvenation (RKVY-RAFTAAR), focusing on boosting farming's economic viability by empowering farmers, reducing risks, and promoting entrepreneurship in agri-business. This was achieved through financial support and fostering the incubation ecosystem. As of 2022-23, RKVY has been revamped into the RKVY Cafeteria Scheme, consolidating various schemes under the Department of Agriculture & Farmers Welfare. These include initiatives such as Soil Health & Fertility, Rainfed Area Development, Paramparagat Krishi Vikas Yojana (PKVY), Per Drop More Crop, Agriculture Mechanization (covering Promotion of Agricultural Mechanization and Management of Crop Residues), Village Haats & GRAAMS, and Crop Diversification Programme. The RKVY Cafeteria Scheme comprises three primary components: the Annual Action Plan (AAP), the Detailed Project Report (DPR), and Administration, Monitoring, and Evaluation, including startup initiatives (DAFW, GoI, 2022).

6.5 Regional Plan (BGREI)

The Bringing Green Revolution to Eastern India (BGREI) is a component of the RKVY that commenced in 2010-11. It operates in seven eastern states: Assam, Bihar, Chhattisgarh, Jharkhand, Odisha, Eastern Uttar Pradesh, and West Bengal. This program aids farmers through cluster demonstrations concerning rice and wheat, seed production and distribution, nutrient management, integrated pest management, training on cropping system-based practices, and the provision of assets such as farm machinery, irrigation tools, site-specific activities, and post-harvest and marketing assistance. Funding is provided by the Government of India to the States, with further allocation to districts managed by the respective state governments. BGREI's objective is to overcome the limitations impacting the productivity of rice-centric cropping systems in eastern India, with a specific emphasis on boosting the productivity of various crops within these systems (PIB, 2021).

6.6 Aspirational District Plan

The Aspirational Districts Programme (ADP), initiated by the Hon'ble Prime Minister in January 2018, seeks to bring rapid transformation to 112 of India's most underdeveloped districts. This program is founded on the principles of Convergence, Collaboration, and Competition, emphasizing the integration of Central and State schemes, cooperation among Central and State-level authorities, and fostering competition among districts through monthly ranking assessments. States are pivotal in driving this initiative, focusing on leveraging each district's strengths, addressing immediate improvement opportunities, and monitoring progress through regular rankings based on 49 Key Performance Indicators (KPIs) across five socio-economic themes: Health & Nutrition, Education, Agriculture & Water Resources, Financial Inclusion & Skill Development, and Infrastructure (NITI Ayog, GoI, 2023). NITI Aayog recently recognized the top-performing aspirational districts in the Agriculture and Water Resources sector for December 2021. Malkangiri in Odisha achieved the highest ranking. Sitamarhi, Bihar, showcased commendable progress in Agriculture and Water Resources under the ADP, leading to the approval of five projects with an estimated cost of 302.69 lakh on March 22, 2022. These projects include the Establishment of a Mushroom Spawn Production Unit and the Establishment of a Custom Hiring Centre for Farm Machinery to enhance agricultural mechanization in Sitamarhi (PIB, 2022). The Agriculture and Water Resource indicators under the Aspirational Districts programme include crucial metrics such as Water Positive investments and Employment, Crop insurance coverage, Increase in critical input supply, Mandi digitization, Price Realization improvement, High-value crop cultivation, Agricultural productivity, Animal healthcare coverage, Artificial insemination services, and Soil Health Card distribution (Aspirational District, MoHMA, GoI, 2019).

6.7 National Food Security Mission

The National Food Security Mission (NFSM) commenced in 2007 and has since evolved to include eight components: NFSM-Rice, NFSM-Wheat, NFSM-Pulses, NFSM-Coarse Cereals (Maize, Barley), NFSM-Sub Mission on Nutri-Cereals, NFSM-Commercial Crops, NFSM Oilseeds and Oil palm, and NFSM-Seed Village Programme. Its goals are to boost rice, wheat, pulses, coarse cereals, and nutri-cereal production sustainably, enhance soil fertility and farm-level economy, and restore farmers' confidence. To achieve these aims, NFSM targets low-productivity districts, adopts cropping system-centered interventions, clusters agro-climatic zones, emphasizes pulse production, promotes improved technologies, ensures timely fund allocation, integrates interventions with district plans, and monitors impact continuously for effectiveness.

6.8 MIDH

The Mission for Integrated Development of Horticulture (MIDH) is a centrally sponsored scheme focusing on the holistic development of the horticulture sector, covering various crops such as fruits, vegetables, root & tuber crops, mushrooms, spices, flowers, aromatic plants, coconut, cashew, cocoa, and bamboo. Under MIDH, the Government of India (GOI) contributes 60% of the total expenditure for development programs in most states, while in the North East and Himalayan regions, this contribution is 90%, with the remaining 40% or 10% coming from the State Governments. The key extension strategies of MIDH include conducting baseline surveys, involving Panchayati Raj Institutions (PRI), creating area-based annual and perspective plans, conducting applied research focused on specific regions and crops, promoting demand-driven production through cluster approaches, ensuring availability of quality seeds and planting materials, implementing technology-driven programs to enhance productivity and quality (such as introducing improved varieties, rejuvenating with improved cultivars, establishing High-Density Plantations, using plastics, promoting bee-keeping for crop pollination), capacity building of farmers and personnel, mechanization, demonstrating latest technologies, managing post-harvest processes and cold chain, developing marketing infrastructure, forming and promoting Farmer Interest Groups (FIGs), Farmer Producer Companies (FPCs), and Farmer Producer Organizations (FPOs), generating, compiling, and analyzing databases with technical support. Additionally, the National Horticulture Mission (NHM) operates as a sub-scheme of MIDH, managed by State Horticulture Missions (SHM) in selected districts across 18 States and 6 Union Territories. Similarly, the National Bamboo Mission, National Horticulture Board (NHB), and Coconut Development Board (CDB) function as 100% centrally sponsored schemes under the ambit of MIDH.

6.9 Commodity Boards (Central Silk Board, Rubber Board, Spices Board, Coffee Board, Tea Board, Tobacco Board, etc.)

The Commodity Boards operating under the Central Ministry of Commerce oversee research and extension activities for specific commodities like Silk, Rubber, Spices, Coffee, Tobacco, and Tea. These Boards have significantly contributed to improving production capacity and supporting farmers by offering necessary facilities. Their functions and objectives encompass various areas:

- Advisory Role: Commodity Boards advise the government on policy matters, including setting quotas and negotiating agreements with foreign countries related to the specific commodities.
- Promotion Activities: They actively promote agricultural commodities by participating in exhibitions, trade fairs, and establishing foreign offices to expand market reach and enhance export opportunities.
- Problem Resolution: Commodity Boards address and resolve issues related to the production, marketing, and distribution of the commodities within their jurisdiction.
- Research and Development: These boards conduct research activities through dedicated research units to innovate and improve production techniques, quality, and marketing strategies for the specific commodities.
- Training Initiatives: Commodity Boards provide training programs and workshops to agriculturists engaged in the production of the designated commodities, aiming to enhance their skills and knowledge.
- Financial Support: Through subsidy schemes and financial assistance programs, Commodity Boards facilitate growers' access to agricultural equipment and resources, promoting efficiency and productivity in commodity production.

7. Public Private Partnership Extension

Public and private entities often have disparate objectives, yet they can achieve mutual benefits through collaborative efforts in Public-Private Partnerships (PPPs). In the realm of Agricultural Extension Management, the concept of PPPs was first introduced in Hoshangabad district, Madhya Pradesh, in 2001. This initiative brought together the Department of Agriculture, Government of Madhya Pradesh, and the Dhanuka Group, facilitated by the National Institute of Agricultural Extension and Management (MANAGE). While the term "private" in PPPs typically refers to the corporate sector, it encompasses a diverse array of organizations, including businesses, community-based groups, and NGOs.

An illustrative example of successful PPP collaboration is seen in Patna, Bihar, where Baidyanath Ayurveda Bhawan Ltd. partnered with growers through the ATMA FIG. This partnership involves the buy-back arrangement for all high-quality herbs and medicinal plants supplied by farmers. Such collaborations exemplify an integrated approach to production and marketing, fostering synergy between public and private entities for the advancement of agricultural practices and economic prosperity.

7.1 Government Regulated Private Extension

The human resource development aspect in the capacity building of extension professionals at the grassroots level involves input dealers who serve as both input providers and advisors to farmers. These private extension professionals play a crucial role as facilitators and require continuous support in terms of emerging technologies, knowledge enhancement, and skill development. Given that they are affiliated with private companies, it's essential for them to adhere to government regulations concerning quality standards and ethical practices, while being integrated into mainstream extension activities. This model, often referred to as a franchise agreement, represents a form of public-private partnership.

Key initiatives in this realm include

- Diploma in Agricultural Extension Service for Input Dealers, organized by ATMA or KVK in a district, with sponsorship and regulation by MANAGE.
- Input Dealers Certificate Course conducted by KVK, sponsored and regulated by NIPHM.
- Self-financed emerging courses for license updates for private dealers, conducted by any HRD institute (public or private), with certificates issued by MANAGE, NIPHM, IIPM, among others.

7.2 Government franchised Private Extension

This extension system involves the government franchising out to the private sector to provide subsidized agricultural inputs and services. The government provides financial support to private organizations to enable the provision of these services at reduced or zero cost, such as government contracting with private agencies for subsidized agro-inputs and agro-services. This model is often referred to as subsidy arrangements or grants, representing a form of public-private partnership in extension services.

7.3 Government franchised Grassroot Group led Extension

Government initiatives that encourage individuals or groups to mobilize group dynamics for their own services and development represent a form of public-private extension known as self-help. For example, the government may facilitate the creation of farmer interest groups and commodity-based groups, providing them with support to access services and information. This approach empowers communities to take ownership of their development by leveraging collective efforts and resources.

Government Mission executed by NGO/FPO/SHG: A case of Millet Mission in Odisha (Gender Equity, Climate Resilience, Nutrition Security)

The Odisha Millet Mission (OMM) is a collaborative effort involving multiple stakeholders led by three partner organizations: the Government of Odisha for policy design and funding, NCDS (Nabakrushna Choudhury Centre for Development Studies) for state/district-level consultations, and WASSAN (Watershed Support Services and Activities Network) NGO for annual and monthly participatory implementation plans, along with research consultancy support from ICAR-IIMR and CFTIR. Launched through a decentralized operational framework, OMM aims to accelerate the production and consumption of millets in Odisha.

The program encompasses several key strategies and initiatives

- Encouraging Household-level Consumption: Awareness campaigns, food festivals, millet recipe events, and street plays are conducted at the village and Gram Panchayat levels in collaboration with Women Self-Help Groups (WSHGs), Anganwadi workers, and local influencers.
- Advancing Decentralized Post-Harvest & Processing Facilities: WSHGs are equipped with threshers and tarpaulins to provide post-harvest services to farmers at the village and GP levels.
- Promoting Value Addition Ventures: Tiffin centers, kiosks, Millet cafes, Millet on Wheels, and Millet Shakti restaurants are encouraged at various levels through WSHGs, Farmer Producer Organizations (FPOs), and other entities.
- Enhancing Millet Crop Productivity: Demonstrations showcasing improved agronomic practices and natural farming techniques are organized, along with establishing custom hiring centers for farming implements.
- Preserving and Promoting Landraces: Participatory varietal trials are conducted with local landraces and improved varieties under technical guidance from institutions like OUAT and IIMR, leading to seed release and multiplication.
- Marketing and Exports: WSHGs and FPOs engage in village-level cleaning and aggregation, supplying bulk quantities to market players and MSMEs, and are registered as export agencies.
- Procurement and Integration of Millets in Government Schemes: The Tribal Development Cooperative Corporation Ltd (TDCCL) serves as

the nodal procurement agency, collaborating with FPOs and WSHGs for millet procurement and distribution through PDS and ICDS.

In FY 2021-22, OMM reached 1.18 lakh farmers across 54,495 hectares, with plans to extend coverage to 1.50 lakh farmers and 81,700 hectares. They established units for processing and value addition, community-managed seed centers, and custom hiring centers. Millet procurement and distribution have been substantial, benefiting farmers and contributing to millet consumption under various government schemes.

7.4 Government Supported Agri-graduates ACABC Extension

The Agriculture Clinic and Agriculture Business Centers (ACABC) scheme, launched in 2002, is designed to empower young rural agriculture graduates who aspire to become agricultural entrepreneurs. This program provides essential training and subsidies to help establish rural service centers, often supported by bank loans. ACABCs offer a wide range of services, including input sales, agricultural advisory, and marketing support, enabling entrepreneurs to earn income and gain prestige as consultants to farmers through Agribusiness Centers. Agriculture graduates interested in setting up Agriclinics or Agribusiness Centers can benefit from free training provided by select institutes. This 45-day course covers topics such as entrepreneurship, business management, and skill enhancement, equipping participants with the necessary knowledge and tools to succeed in their ventures. Additionally, bank loans are accessible for Agriclinics and Agribusiness Centers, with increased project cost ceilings and flexible repayment options. Under this scheme, individual projects can receive subsidies for up to Rs. 20 lakhs (or Rs. 25 lakhs for highly successful individual projects), while group projects can receive subsidies of up to Rs. 100 lakhs. The terms of the loan, including the rate of interest, margin, and security requirements, are determined by the respective bank, following guidelines set by the Reserve Bank of India (RBI) to ensure financial stability and sustainability.

7.5 Para professional led Extension (Kissan Mitra, Gopal Mitra, Matsya Mitra etc.)

The system described is known as a franchise agreement extension system, where the government authorizes private entities to provide specific services. These services are directly accessed by users through local grassroots professionals who are trained and monitored by the government. For instance, paravets offer services like artificial insemination (AI) and animal vaccinations, while other para-extension workers like Kisan Mitra and Matschya Mitra provide agricultural services such as soil testing and fisheries extension. Para-extension workers play a crucial role in complementing public extension services in a cost-effective manner and overcoming the challenges posed by absentee public extension officials. They receive training and capacity building funded by the government and are initially compensated with an honorarium. This honorarium is channeled through Farmer Groups or Farmer Organizations, ensuring accountability and quality service delivery to the client group.

8. Privatized Extension

8.1 Input Agency Extension

The agricultural sector relies significantly on the approximately 2.82 lakh agro-input dealers operating across rural areas in India, serving as crucial sources of information and support for farmers. Apart from providing inputs and advisory services, these dealers are increasingly playing roles in extending credit and facilitating product sales post-harvest. Corporates in the agricultural sector are actively engaging with farmers by conducting demonstrations, field days showcasing emerging technologies, sponsoring seminars, and supporting farmer training programs. Companies like Bayer, Syngenta, Pioneer, Monsanto, and Nuziveedu Seeds are notable contributors to farmer education and technology dissemination. In high-value crop segments such as vegetables and flowers, companies often provide end-to-end extension support, encompassing various stages from cultivation to marketing. Fertilizer companies like Indian Farmers Fertilizer Cooperative Limited (IFFCO) and Krishak Bharati Cooperative (KRIBHCO) are also heavily involved in extension activities. They conduct farmers' meetings, organize crop seminars, host exhibitions and fairs, offer soil testing facilities, and even implement village adoption programs. Collaborations with government departments and Krishi Vigyan Kendras (KVKs) further enhance the reach and impact of these extension activities by leveraging expert services and resources.

8.2 Consultancy Service Sector Extension

Tata Chemicals Limited, a prominent agricultural chemical company, initiated the Tata Kisan Kendras (TTKs) in 1998 to provide farmers with a holistic range of inputs and services. This initiative evolved into the Tata Kisan Sansars (TKS), comprising nearly 600 farmer resource centers that cater to over 3.5 million farmers across 22,000 villages in northern and eastern India. The TKS operates through a hub-and-spoke model, where each center acts as a franchised retail outlet and a problem-solving hub for around 30-40 neighboring villages. These centers receive support from approximately 30 resource centers known as Tata Krishi Vikas Kendras (TKVK), with each TKVK supervising around 17-18 TKS centers. The TKVKs are staffed with over 60 agronomists who provide expert guidance on crop management and farming issues, ensuring comprehensive support to the farming community.

8.3 Financial Agencies extension

NABKISAN Finance Limited (NKFL) is recognized as a Non-Banking Finance Company (NBFC) by the Reserve Bank of India (RBI) with a primary objective of providing financial assistance to agricultural, allied, and rural non-farm enterprises. NKFL plays a crucial role in supporting livelihoods and income-generating ventures by extending credit to various entities such as Self-Help Groups (SHGs), Joint Liability Groups (JLGs), and Farmers' Producers' Organizations (FPOs). As of December 2022, NABKISAN has approved over 1800 loans to FPOs, establishing itself as a key player in the FPO lending landscape. The company operates across 21 states, including prominent eastern Indian states like Odisha, Bihar, Jharkhand, West Bengal, and Chhattisgarh. In its bulk lending segment, NABKISAN prioritizes lending to NBFCs operating in rural areas, FPOs, agriculture and allied activities, value chain financing, and MSMEs engaged in income-generating endeavors, thereby contributing significantly to rural financial inclusion and economic empowerment (NABARAD, 2023)

8.4 Market Led Extension

Market-led extension strategies are crucial in the marketing of perishable fruits like litchi, which have a short shelf life and require specific storage conditions. Litchi growers often face challenges in self-marketing due to these factors. The predominant marketing channels for litchi include pre-harvest contractors, direct sales to wholesalers, and village-level agents/commission agents.

1. Pre-harvest contractors: This channel involves contractors taking control of orchards, purchasing litchi produce from farmers, and paying a predetermined amount after sales. While this method provides convenience to farmers, contractors often profit significantly, and farmers may receive lower prices for their produce.
2. Direct sales to wholesalers: Some farmers opt to sell their litchi directly to wholesalers, who then distribute the fruit to retailers and consumers. This channel allows farmers to negotiate prices directly, potentially increasing their earnings.
3. Village-level agents/commission agents: Farmers can also engage agents at the village level or commission agents for direct procurement and marketing of litchi. This channel provides flexibility but may involve additional costs and complexities.

To enhance litchi fruit income, farmers can explore alternative marketing strategies such as small-scale direct marketing or forming farmer cooperatives. Direct marketing enables farmers to sell directly to consumers or local markets, bypassing intermediaries and potentially fetching better prices. Farmer cooperatives allow for collective marketing, pooling resources, and negotiating better deals with buyers. Improving infrastructure like cold storage facilities and efficient logistics is essential for maintaining litchi quality and extending its market reach. Policy support is crucial in facilitating infrastructure development, providing training in marketing strategies, and creating conducive environments for farmers to engage in direct marketing or cooperative ventures.

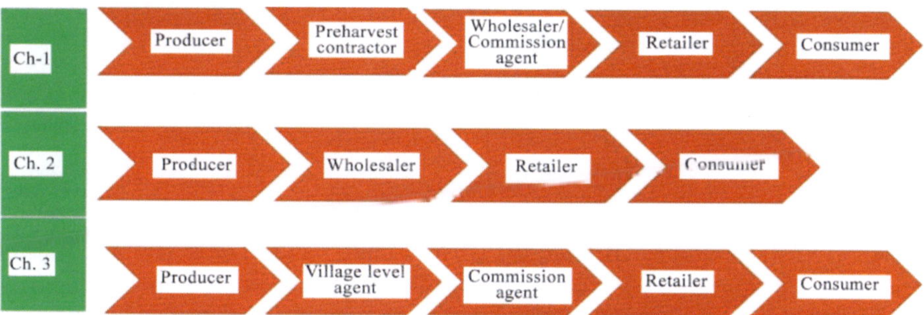

Fig. 5: Predominant marketing channels for litchi in Bihar (Adapted from Kumar *et al,* 2017)

Moreover, advancements in technology such as e-marketing and modern logistics can revolutionize litchi marketing. E-marketing offers advantages like traceability and product quality assurance, transforming the conventional seller-seeking-buyer model into one where buyers actively seek out the product. Innovative treatments like the one developed by BARC, extending litchi's shelf life, further enhance market opportunities for litchi growers, especially in distant markets. Collaboration between research institutions and agricultural organizations can drive such technological advancements and benefit farmers in the long run.

8.5 Contract Farming

Frito-Lay, a division of PepsiCo specializing in food products, stands as the largest potato procurement entity in collaboration with contract farming. Approximately 50% of their total potato procurement is sourced through contract farming arrangements. PepsiCo has forged partnerships with over 12,000 farmers across more than 6400 hectares of land in several states such as Punjab, Uttar Pradesh, Karnataka, Bihar, West Bengal, Gujarat, and Maharashtra to ensure a steady potato supply. This collaboration benefits both farmers and the company, as it fosters adherence to Good Agricultural Practices through pre-agreed market commitments, ensuring the company receives high-quality produce. In West Bengal, Bayer Crop Science has joined forces as a comprehensive crop solution provider in the Food chain partnership project associated with this contract farming initiative. As part of this project, a tailored plant protection kit known as "5P" (Production, Protection, Program Monitoring, Passports, and Post-Harvest) is provided to participating farmers. This kit includes guidance and monitoring throughout the crop season. Farmers are equipped with a passport developed by Bayer Crop Science, serving as a traceability tool and a record-keeping document for their farming activities. The passport not only enables traceability but also allows farmers to conduct cost-benefit analyses of their operations, contributing to enhanced efficiency and sustainability (Bayer.com, 2012).

8.6 NGO/CSO Led Extension (PRADAAN, IFAD, BAIF, WASSAN, Local NGOs

Several prominent non-governmental organizations (NGOs) are active across various states in India, such as the Bharatiya Agro-Industries Federation (BAIF), Professional Assistance for Development Action (PRADAN), Action for Food Production (AFPRO), and the Foundation for Ecological Security (FES) (Gulati, 2018).

PRADAN, founded in 1983, is a prominent Indian non-governmental organization (NGO) with a strong focus on poverty alleviation through a people-centric approach. Instead of following a prescriptive path, PRADAN emphasizes deploying high-quality human resources in villages to facilitate comprehensive change. The organization collaborates with various partners in eastern India, including the Bihar Rural Livelihood Promotion Society, Chhattisgarh Grameen Aajeevika Samvardhan Samiti, Directorate of Horticulture, Odisha, Jharkhand State Livelihoods Promotion Society, West Bengal Panchayati Raj Department, The Agriculture Promotion and Investment Corporation of Odisha Limited, and West Bengal State Rural Livelihood Mission, among others. PRADAN's efforts are focused on assisting ultra-poor communities in establishing their own financial base through micro-savings and accessing additional capital from banks, as these communities often lack reliable sources of finance. The organization works on community mobilization of women's Self-Help Groups (SHGs), enhancing access to finance, livelihood finance, increasing income, and empowering women in rural villages. PRADAN's initiatives impact approximately 10 million people across 9,127 villages in 40 of the poorest districts in seven eastern Indian states, as outlined in their Annual Report for 2021-22.

BAIF, on the other hand, primarily concentrates on livestock development, water resource management, environmental conservation, and livelihood enhancement. Operating through 9 associate organizations, BAIF's reach extends to 16 states across India.

The Syngenta Foundation India (SFI) supports sustainable agricultural projects aimed at improving long-term productivity and income for farmers. Key areas of focus include water conservation, developing varieties suitable for local conditions, and providing farmers with access to relevant information. SFI collaborates with the Syngenta Foundation for Sustainable Agriculture and local NGOs to achieve its objectives.

Grameen Vikas Trust is dedicated to empowering rural communities in resource-poor areas to enhance their livelihoods. Operating in over 150 villages across eight states, the organization's activities span natural resource management, watershed development, agriculture, livelihood improvement, institutional development, women empowerment, labor support, and microenterprise development (Rasheed, 2012).

8.7 Private Agricultural Educational Institutions led Extension

Centurion University, located in the Gajapati district of Odisha and home to a College of Agriculture, has initiated several outreach programs to engage with communities and address their holistic needs. One of their key initiatives is the frontline extension program conducted in collaboration with public research and educational institutes in adopted villages. This program involves agricultural and allied students and faculty members conducting baseline surveys to identify community needs and design programs that cater to these aspirations through a multidisciplinary approach. One notable program conducted by Centurion University is the Nutritional Enrichment and Capacity Development program. Through this initiative, dairy and mushroom products from the university are provided to the community at reasonable costs. Additionally, drumstick and other vegetable seedlings are distributed to women farmers to promote the development of nutria-smart kitchen gardens. The university also organizes vocational training sponsored by ATMA, which spans six days and benefits communities. Moreover, Centurion University is committed to training 1.0 lakh farmers in Odisha under Project ATAL: Recognition of Prior Learn (RPL). This project, implemented across 18 districts in four zones of Odisha, particularly targets vulnerable and weaker sections of society, aiming to enhance their agricultural skills and knowledge.

Similarly, GIET University in Rayagada offers a Rural Agricultural Work Experience (RAWE) program. As part of this program, graduating students undergo a 14-week block attachment training and participate in the Krushi

Unnata Sahajogi (KUS) program under the Government of Odisha. Through KUS, students gain practical experience and insights by working alongside department officials and farmers, learning about various agricultural schemes and acquiring essential skills. They use a mobile app called the KUS app to collect data and maintain attendance records, which helps in decision-making at the state level, particularly during critical cropping seasons. The program has proven beneficial in identifying and addressing issues such as Brown Plant Hopper (BPH) attacks in rice, allowing for proactive measures by the state department based on students' observations and data collected through the app.

9. Cyber Extension

Public sector ICT initiatives

9.1 Farmers Portal: The Farmers Portal serves as a comprehensive platform providing information on crop insurance, storage techniques, crop advisories, extension activities, seeds, pesticides, farm machinery, fertilizers, and market prices. It also offers details and guidelines regarding various agricultural schemes.

9.2 m-Kisan: This unified platform facilitates the delivery of targeted text and voice messages in local languages to farmers, sent by officials and scientists. Technical weather-based advisories are provided by Krishi Vigyan Kendras (KVKs), State Agricultural Universities (SAUs), and state agricultural departments to registered farmers' mobiles, offering information on preparedness and contingency measures for events like cyclones, storms, rainfall, or disease and pest attacks.

9.3 Kisan Call Centre: A toll-free number (1800-180-1551) accessible via mobile and landline phones, the Kisan Call Centre provides agricultural advisory services in local languages. Farmers can seek information and advice on various agriculture-related issues through specialist consultations and callback services.

9.4 Kissan Sarathi: This application facilitates farmers' access to timely and accurate information in their preferred language. It allows officials to monitor daily operations, including farmer registration, live calls, message exchanges, advisory services, and pending tasks. Farmers can engage with scientists at KVKs through this platform to receive personalized guidance on agriculture-related topics.

9.5 Kisan TV Channel: DD Kisan, a 24-hour TV channel dedicated to agriculture, provides real-time information in regional languages on inputs, farming techniques, water conservation, and more, specifically targeting farmers. Additionally, various Doordarshan regional language TV shows feature dedicated agricultural segments, such as ETV Annadatta and DD Odia Pallishree.

9.6 All India Radio (AIR) Farm & Home Unit: The Farm & Home unit of AIR broadcasts programs like "Kishanvani" to raise awareness among farmers about promoting the cultivation of oilseeds and pulses in rice fallow areas, aligning with government initiatives like Bringing Green Revolution to Eastern India (BGREI). The "Kisanvani" program offers daily news, market updates, weather information, interviews with farmers, and news from around the country and abroad. It engages farmers through interactive content and provides valuable information on agriculture and market prices.

Portals and Apps Developed by State Departments

- e-PACS: A mobile and web-based solution facilitating paddy procurement through Primary Agriculture Credit Societies (PACS) and Vyapaar Mandals in Bihar.
- Various mobile apps by the Government of India: These apps provide customized information on agriculture, market prices, weather forecasts, crop insurance, and other relevant topics.
- Agriextension, Kalia portal, and GOSUGAM portal: These platforms aim to transform farmers' lives by leveraging technology to deliver timely benefits and services under farmer-centric schemes.

During the COVID-19 crisis, agripreneurs in Eastern India leveraged alternative communication methods amid travel restrictions, embracing various cyber extension methods to support farmers effectively. They utilized increased mobile phone usage among farmers for communication, leveraging mobile calling and the WhatsApp platform extensively for online interactions and information dissemination. Despite travel limitations, agripreneurs found ways to visit farmers' fields personally, offering advice, raising awareness about ICT tools, and establishing direct connections. They also made use of various ICT tools and digital platforms for information delivery and interactive content creation, complemented by print media to distribute educational materials. Agripreneurs provided a range of services during the pandemic, including crop-based advisory, input supply, counseling, pest diagnosis, government scheme awareness, COVID-19 education, and support for returning migrants, ensuring comprehensive support for farmers' well-being and agricultural activities.

10. Gender Sensitized Extension

10.1 Women Para Extension Worker Model

The Gender Sensitized Extension approach involves innovative strategies to empower women in agriculture and allied activities. One such model is the

Women Para Extension Worker Model, developed by the Central Institute for Women in Agriculture, Bhubaneswar. This model emphasizes the role of Village level Para Extension Workers (VPEWs) to promote gender sensitivity, leadership development among women in farming, and location-specific technology adoption. Additionally, Mission Shakti in Odisha is a pivotal government initiative that focuses on empowering women through economic activities, credit access, and market connections. This flagship program has organized around 6 lakh groups comprising 70 lakh women across the state.

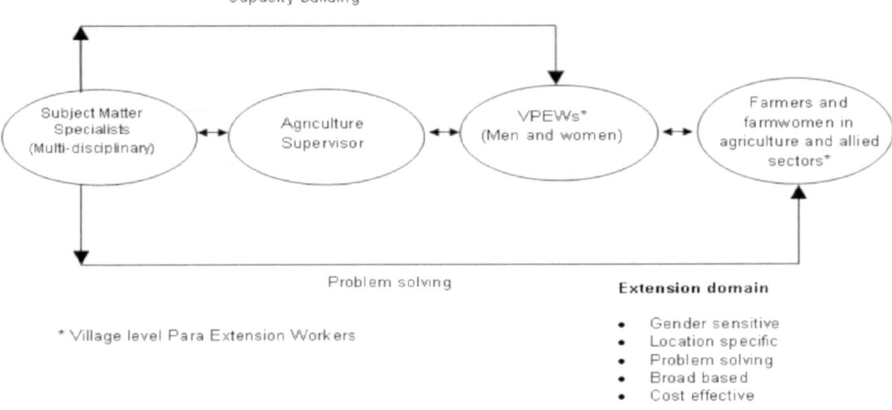

Fig. 6: Schematic Gender Sensitive Extension Model (Adapted from CIWA, 2007)

10.2. Mission Shakti also provides various support measures such as training, on-farm extension support for pisciculture and horticulture, livelihood assistance, and skill-based training for income-generating activities

10.3. ORMAS (Odisha Rural Development and Marketing Society) complements these efforts by promoting livelihood initiatives, forming new Women Producer Groups (PGs) and Self-help groups, and developing Agricultural Production Clusters and Mission Shakti Bazaar Outlets. The target audience for gender-sensitive extension services includes a wide range of stakeholders, from farmers and informal sector workers to individuals involved in food processing and off-farm businesses, with a focus on empowering women across various sectors and activities.

11. Group Led Extension

This initiative aims to encourage the establishment and mobilization of Farmer Interest Groups (FIGs), Farmers Co-operatives, and Self-Help Groups with the support of NGOs. The approach to group extension will shift from a top-down to a bottom-up method in technology transfer, where FIGs generate demand for information and technology, and extension workers respond to this demand, fostering a participatory problem-solving approach. This group-based extension is aligned with rural credit delivery, Self-Help Groups, Water User Associations, and cooperatives, as emphasized by SAMETI Jharkhand (2023).

User groups, including farmer interest groups, farmer clubs, commodity groups, women farmer groups, and special interest groups, play a pivotal role in agricultural extension. The government's objective is to promote farmer organizations at the village level, empowering them in addressing challenges effectively. An exemplar of a successful farmer organization is the Grape Growers Association of Maharashtra (MRDBS), which provides essential services and support to grape producers, contributing significantly to the international recognition of Maharashtra's grapes. Similarly, the United Planters Association of Southern India (UPASI) supports tea, coffee, rubber, and cardamom growers, engaging in research, extension, and representation of grower interests. Several states, notably Kerala, have initiated the formation of farmer groups, leading to commodity-specific groups focused on crops like paddy, coconut, and pepper. Promoting farmer self-help groups is a crucial component of projects like the Kerala Horticultural Development Programme (KHDP) and Uttar Pradesh Diversified Agricultural Support Project (UPDASP). Self-Help Groups (SHGs) have evolved into institutions such as the Vegetable and Fruit Promotion Council, Kerala (VFPCK) (Rasheed, 2012). The formation of farmer interest groups and their federation at the block and district levels is a key strategy in the ATMA extension approach, aimed at enhancing productivity and income while reducing government dependence. However, many of these groups remain inactive due to a lack of sustained efforts.

11.1 Cooperative Society-led Extension

Agricultural cooperative societies involved in input supply, credit distribution, and processing cooperatives such as Sugar Cooperatives and Dairy cooperatives have flourished due to their consumer products. Conversely, large cooperatives like KRIBHCO and IFFCO have succeeded in marketing fertilizers to both farmer members and external markets as profitable business enterprises. An assessment of Primary Agricultural Cooperative Societies (PACS) indicates that 19% of them have the potential to be viable, 49% are operating profitably, and 45% cater to small farmers who borrow from these societies. However, approximately 54% of the total staff, amounting to 93,432 individuals out of a total of 172, 287, lack training, posing a significant obstacle to the sustainability of cooperatives. Despite these challenges, there is an 85% coverage of villages, with 51,939 societies having storage structures (go-downs) (Sahoo *et al.*, 2020).

11.2 FPO-led Extension

The establishment of Farmer Producer Organizations (FPOs) plays a crucial role in achieving economies of scale across various agricultural activities. The Government of India has actively promoted the creation of FPOs through different programs and schemes over the past decade, with organizations like SFAC and NABARD offering specific schemes for this purpose. These Producer Companies operate based on mutual assistance and patronage principles, combining the advantages of both cooperative and corporate sectors to benefit small and marginal farmers. Members of FPOs experience reduced input costs, improved market access, and enhanced bargaining power due to aggregation. Additionally, the advisory and value-added services provided by FPOs contribute to better decision-making among farmers (Kumar *et al.*, 2022).

11.3 Demonstration, Field Day, Training, Meetings: These group-oriented approaches are utilized to sensitize, inspire, and convince action among farmers. They are commonly used by change agents due to their significant impact on programs.

- Demonstration is a widely adopted tool to showcase improved technologies on farmers' fields, facilitating hands-on learning and convincing farmers of the benefits of new technologies.
- Field day is the culmination of result demonstration, where farmers gather to see technologies demonstrated on fields, conveying success or lacuna in demonstration to fellow farmers.
- Training is an active learning process where farmers acquire specific skills, knowledge, and attitudes towards new learnings, enhancing competency for specific tasks.
- Meetings facilitate group discussions on problems, solutions, and new schemes, leveraging group dynamics for faster adoption due to peer pressure, and incorporating stakeholders' inputs in policy and program design.

12. Climate Smart Extension

The National Innovation on Climate Resilient Agriculture (NICRA) scheme is a focused initiative aimed at developing and promoting climate-resilient technologies in agriculture to address vulnerable areas in India. Recent updates include efforts to create solutions for extreme weather conditions such as droughts, floods, frost, and heat waves. The strategic research programs have been carried out involving 21 Institutes of the ICAR. The scheme comprises four major components:

- Strategic research on adaptation and mitigation.
- Technology demonstration in 100 vulnerable districts to address current climate variability.
- Capacity building to enhance knowledge and skills among stakeholders.
- Sponsored competitive research to fill critical gaps in climate-resilient agriculture.

Climate-smart agricultural (CSA) practices encompass a wide range of strategies that address crop, soil, and water management. These practices range from tactical decisions like optimizing sowing times and crop maturity to managing nutrients, water, and pest control. They also include conservation tillage techniques and strategic choices such as crop selection, rotations, cropping patterns, and investments in irrigation infrastructure. Field studies help identify effective "climate-smart" practices at specific locations, while crop and cropping system simulation models allow for ex-ante evaluations to predict how these practices will perform under various climate scenarios before implementation.

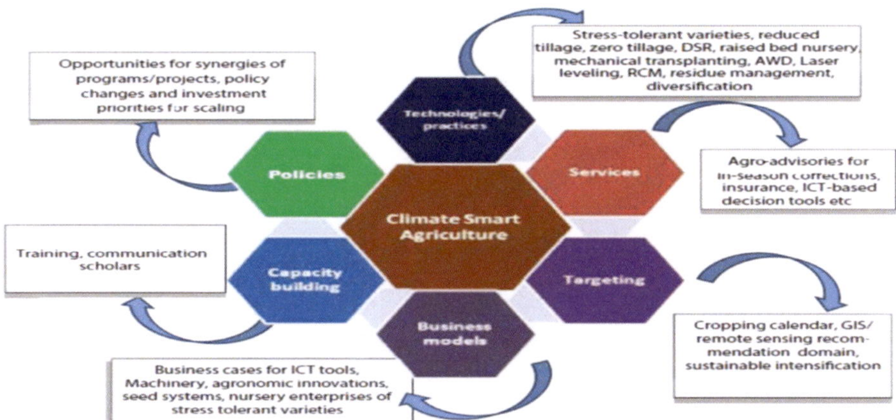

Fig. 7: Climate Smart Extension Strategies for the vulnerable areas in Odisha (Adapted from Sharma *et al.*, 2020)

The Climate Smart Extension Strategies in vulnerable areas of Odisha include various initiatives:

a) Seed Village and Multiplication Programs: Encouraging the cultivation and multiplication of climate-resilient crop varieties through seed villages to ensure a sustainable seed supply.

b) Demonstrations through Farmers' Producers Organizations (FPOs): Promoting climate-resilient practices and technologies by collaborating with NGOs, CSOs, and community institutions to showcase these practices to farmers.

c) **Crop Contingency Plans:** Working with adaptive and progressive farmers to develop crop contingency plans that reduce climate-related risks.

d) **Crop Diversification:** Demonstrating the benefits of crop diversification, moving from paddy to other suitable crops according to land classification, to minimize the risk associated with over-dependence on a single crop.

e) **Capacity Building and Knowledge Dissemination:** Initiating programs on climate change and climate-resilient practices, often through Farmers' Field Schools (FFS) that combine classroom learning with practical demonstrations.

f) **Model Climate-Smart Villages:** Piloting climate-smart and climate-resilient villages in different agro-ecological zones in Odisha, providing end-to-end support, including technology transfer, infrastructure, input provision, weather advisory services, and farmer understanding and confidence-building.

g) **Integration with Existing Government Schemes:** Coordinating with existing government programs related to technology transfer, agricultural inputs, implements, renewable energy, and water conservation for more effective climate-resilient agriculture interventions.

h) **Technical Support:** Collaborating with KVKs, Agriculture Department officials, and other agriculture resource persons to support sustainable and climate-smart agriculture.

i) **Water Budgeting:** Conducting water budgeting exercises at the village level during crop planning to manage water resources efficiently for various purposes.

j) **Partnerships:** Collaborating with institutions such as the IRRI, Philippines; NRRI, Cuttack; OUAT, Bhubaneswar; and the Department of Agriculture and Farmers' Empowerment, Govt. of Odisha to upscale the best climate-resilient practices, crops, and technologies.

13. Mass Approaches

13.1 Agriculture Fairs and Exhibitions

This mass extension approach brings together all stakeholders in one place, showcasing their technologies, expertise, literature, innovations, and attractive exhibits to capture the attention of visitors. This collaborative platform connects public, private, NGOs, SHGs, FPOs, agri-preneurs, corporates, insurance agencies, new startups, farm machinery providers, live demonstrations, folk dances, folk songs, cultural artifacts, etc., making the event lively and informative within a limited period. Farmers can directly interact with experts at various stalls, getting immediate solutions to a wide range of problems. Additionally, there are separate sessions for farmer-scientist interactions, successful farmer awards ceremonies, and farmers' experience-sharing sessions, all of which inspire farmers to adopt improved scientific technologies. The Pusa Krishi Mela serves as a central event, with states organizing one or two state-level and regional Krushi Melas annually. District and block-level melas are also increasingly organized, often initiated by specific departments like horticulture, watershed, NAFED, ORMAS departments, etc., reflecting the success of this approach.

13.2 Awareness, Campaign, Diagnostic Visit, Surveillance

These mass approaches aim to raise awareness and sensitize a larger section of farmers within a limited time period, focusing on timely, theme-based, and season-specific initiatives.

a) Awareness Programs: These programs use various tools such as banners, posters, leaflets, literature, audio-visual aids, sound systems, and mobile vans equipped with sounds and banners to make a larger section of the population aware of important issues such as schemes, beneficiary selection, deadlines, input availability, market openings, and procurement token issuance.

b) Campaigns: Similar to awareness programs, campaigns like seed treatment, Integrated Pest Management (IPM), Soil Health Card distribution, natural farming promotion, water conservation, and nutritional kitchen gardens are based on emerging themes identified by extension agents. These campaigns use mobile vans and "Krushi Ratha" (agriculture chariots) equipped with sound systems to showcase technologies. For instance, soil health camps in Odisha use mobile vans for soil testing, and live vaccination programs are conducted on the spot by veterinary departments during animal health camps.

c) Diagnostic Visits: In case of emergencies such as widespread insect pest attacks or crop diseases affecting many farmers locally or at the district level, a multidisciplinary team comprising experts from State Agricultural Universities (SAUs) and department officials conducts diagnostic visits to fields. They diagnose the problem on-site and provide instant solutions. These teams also engage with the media to communicate problems and solutions in the local language, reaching a larger audience. This surveillance and advisory dissemination continue throughout peak problem periods and are shared through mass media methods like print, TV, radio, portals, etc.

d) Pest Surveillance: Pest surveillance is crucial for timely and effective Integrated Pest Management (IPM). By using pest surveillance technology, insects, diseases, predators, and parasites are identified and counted. This information helps in managing pests and diseases through eco-friendly IPM systems. Odisha has implemented the e-Pest Surveillance Programme since 2010, allowing real-time pest surveillance updates from Village Agricultural Workers (VAWs) to experts from SAUs and Krishi Vigyan Kendras (KVKs) and Joint Directors (Plant Protection). This system provides instant pest advisories in the Odia language to farmers via mobile phones, with a dashboard for monitoring pest/disease situations across the state, collecting expert advisories online during critical periods, and reflecting them on the website immediately.

14. Farmer to Farmer Extension

The Farmer-to-Farmer Extension (F2FE) approach revolves around proactive farmers who willingly adopt and experiment with new agricultural technologies or seed varieties. These farmer trainers, often lacking formal education, enhance their knowledge through training, experimentation, and practical experience. They then share this knowledge with other farmers within their social networks, essentially functioning as extension workers. These identified farmer trainers can be chosen or appointed, with or without financial incentives, to facilitate the diffusion of technology in specific regions. Initially, these farmers closely collaborate with experts to explore technological options, develop skills related to technology, experiment, and subsequently share the acquired knowledge with other farmers in their social circles. The F2FE approach empowers these farmer leaders to be catalysts for change through close collaboration with change agents from various public and private agencies, thereby enhancing the diffusion of technology among farmers. Identifying and collaborating with these types of farmers to work within the extension system is crucial. This approach has the potential to enhance the flow of feedback from farmers to extension staff, ultimately complementing formal extension services in disseminating agricultural technologies and building farmers' capacities. Researchers worldwide use various terms for such community leaders, including progressive farmers, innovative farmers, and certified farmers.

a) Progressive farmer: This farmer has opinion leadership in the community and maintains good contact with extension agencies, serving as a role model for other farmers and often utilized in extension program interventions in local communities.

b) Innovative farmer: This farmer constantly seeks innovations within their community or develops creative solutions to problems. Innovations should be validated and patented by extension agencies to recognize the farmer's efforts.

c) Certified farmer: These are farmers trained by change agents such as departments, SAMETI, KVK, SAUs, private HRD institutions, etc., who have expertise in specific fields and can complement change agents in local extension work.

d) Progressive farmers' clique: This refers to a group of interested progressive farmers who take the initiative in any extension program by change agents. Their close coordination and commitment make their work more effective, especially in convincing other farmers to adopt new practices.

In Farmer-to-Farmer extension methods, lead farmers are trained on the approach, including technology, tools, and methods. They are pioneers who implement and test the approach on their own fields initially and then share their knowledge and skills with fellow farmers through model farms, community networks, practical training sessions, and similar means.

In farmer-trainer programs, extension strategies include:

- Involvement of farmers and local institutions in selecting, monitoring, and evaluating farmer-trainers.
- Farmer-trainers being integral parts of the community, communicating in local languages and understanding local cultures and practices.
- Selection of farmer-trainers based not just on farming expertise but also on effective communication skills.
- Strong linkages and support from development agents, complementing existing extension systems.
- Promotion of gender inclusivity by providing opportunities for women to become farmer-trainers.
- Providing suitable reference materials to farmer-trainers to enhance their effectiveness in disseminating knowledge within the community.

Conclusion

The agricultural extension in Eastern India is pivotal for driving sustainable agricultural development. Despite significant strides made through the green revolution and modern extension services, there remains a substantial disparity in agronomic and socio-economic outcomes between different regions, necessitating a strategic reevaluation of extension strategies and approaches.

The following key recommendations are imperative to bridge these gaps and unlock the full potential of agricultural extension in Eastern India:

1. Strategic Resource Allocation: Increase the allocation of agricultural GDP towards extension and training, with a focused emphasis on sectors such as livestock, fisheries, and allied industries. This strategic realignment will enable a more comprehensive and targeted approach to agricultural development, addressing the specific needs of the region.
2. Collaborative Synergies: Foster stronger collaboration and synergies among public research institutions, civil societies, and government ministries. Facilitate seamless knowledge sharing, technology transfer, and capacity building through collaborative platforms and networks. This collaborative approach will amplify the impact of extension efforts and ensure a more holistic and inclusive agricultural development process.
3. Empowerment of Extension Providers: Define clear roles, responsibilities, and capacities for extension service providers. Establish mechanisms for quality certification and continuous professional development to enhance the credibility and effectiveness of extension services delivered to farmers. Empowering extension providers is crucial for ensuring tailored and impactful extension interventions.
4. Digital Transformation: Embrace digital innovation and leverage digital platforms and tools for information dissemination, technology adoption, and best practices sharing. Digital solutions can enhance outreach, accessibility, and efficiency in delivering extension services, particularly in remote and underserved areas of Eastern India.

By implementing these robust recommendations, Eastern India can transform its agricultural extension ecosystem into a more resilient, efficient, and sustainable model. This will not only bridge existing disparities but also pave the way for inclusive and holistic agricultural development, benefiting farmers and rural communities across the region.

References

AESA. (2023). Retrieved from https://www.aesanetwork.org/blog-191-the-continuing-relevance-of-all-india-radio-for-agricultural-development-in-the-digital-era/

Agriclinics.net. (2023). AGRICLINICS & AGRIBUSINESS CENTRES Better Farming by Every Farmer. Retrieved from https://www.agriclinics.net/AboutScheme.aspx

Anonymous (2023). Retrieved from https://atariz1.icar.gov.in/ATIC.aspx

APICOL, DAFE, GoO. (2023). Retrieved from https://apicol.nic.in/information?id=1

Aspirational District. MoHA. GoI. 2019. https://www.aspirationaldistricts.in/wp-content/uploads/2019/02/Asipirational_District_Complete_Booklet_5-2-19_B.pdf

Atari, Kolkata. 2023. Retrieved from https://www.atarikolkata.org/wp-content/uploads/2023/04/TSP_2021.pdf

Bayer.com(2012) Retrieved from https://www.bayer.com/sites/default/files/FCP_the_indian_potato_project_EN_screen_RGB.pdf

Bhavani, G., Sreenivasulu, M., Naik, R. V., Reddy, M. J. M., Darekar, A. S. and Reddy, A. A. (2022). Impact assessment of seed village programme by using difference in difference (DiD) approach in Telangana, India. Sustainability, 14(15): 9543.

Caritas India. 2021. Documentation of Traditional Climate Resilient Crops and Agricultural prac¬tices in the selected blocks of Odisha. Retrieved from https://caritasindia.org/GlobalProgramIndia/wp-content/uploads/2022/03/Documentation-of-Traditional-Climate-Resilient-Crops-and-Agricultural-practices-in-the-selected-blocks-of-Odisha.pdf

CIWA,2007. Retrieved from https://icar-ciwa.org.in/gks/index.php/wft/100-vpew

CRISP (2023). Retrieved from http://reachingruralwomen.org/gendersens.htm

Csisa.org.in (2023). Retrieved from https://csisa.org/csisa-locations/csisa-india/

Cutm.ac.in. (2022) Retrieved from https://cutm.ac.in/wp-content/uploads/2022/booklet/University-Extension-Activities.pdf

DAFE (2022). Schemes, Department of Agriculture & Farmers Empowerment, Government of Odisha. Retrieved from https://agri.odisha.gov.in/sites/default/files/2022-10/SCHEMES.pdf

DAFW, GoI. 2022. https://agricoop.gov.in/sites/default/files/rkvy_inro.pdf

Dhar, A. (2016). Retrieved from https://csisa.org/tag/indian-government/

DoE, DAFW, MoA, GoI, (2018) https://extensionreforms.dacnet.nic.in/PDF/atmaguid23814.pdf

DoE, DAFW, MOA, GoI. 2018. Retrieved from https://extensionreforms.dacnet.nic.in/PDF/atmaguid23814.pdf

EEI, Jorhat. (2019). Retrieved from http://www.eei-ner.org/wp-content/uploads/2019/09/Handout-NMET-EMS.pdf

Epestodisha.nic.in (2023). Retrieved from https://www.epestodisha.nic.in/

FARD (2018). Annual Activity Report. Fisheries & Animal Resources Development Department. Retrieved fromhttps://odishaahvs.nic.in/upload/files/Annual%20Activity%20Report%20of%20ARD%20Sector%202017%20-%2018_10_18_06am95da105629fe20fc08864796b3ad35e0.pdf

Gates foundation. 2023. https://www.gatesfoundation.org/our-work/places/india/agricultural-development

GFRAS. (2012). Retrieved from file:///C:/Users/LENOVO/Desktop/Book%20chapter%20on%20Extension%20in%20Eastern%20India/gfras_newextensionist_positionpaper.pdf

Giet.edu. (2023) Retrieved from https://www.giet.edu/courses/b-sc-hons-agriculture/rawe-kus-programme/

Gulati, A., Sharma, P., Samantara, A. and Terway, P. (2018). Agriculture extension system in In¬dia: Review of current status, trends and the way forward. Retrieved from https://tile.loc.gov/storage-services/service/gdc/gdcovop/2018305074/2018305074.pdf

ICAR-ATARI, Hyderabad, 2023. Retrieved from https://atari-hyderabad.icar.gov.in/atarihyderabad/mgmg?lang=en

ICAR-CIWA. 2020. Retrieved from https://icar-ciwa.org.in/pdf/SCSP/SCSP-Annual%20Report-2019-20.pdf

ICAR-CRIJAF. 2022. Retrieved from https://crijaf.icar.gov.in/SideLinks/GovtNsponserdPrjcts/scspReport.pdf

ICAR-FFP. (2017) ffp.icar.gov.in Retrieved from https://ffp.icar.gov.in/Homepage.aspx

ICAR-IIWM. 2022. Retrieved from http://www.iiwm.res.in/news/SCSP_Sep2022.pdf
ICAR-NRRI. (2023) Retrieved from https://icar-nrri.in/tribal-sub-plan/
ICAR-NRRI. 2023. Retrieved from https://icar-nrri.in/training-programme-under-scsp/
ICAR-NRRI. In (2023). Retrieved from https://icar-nrri.in/abi-unit/
IIPM, (2023). Retrieved from https://iipmb.edu.in/about-us/
Jana, (2016). Role of Extension Education Institute (EEI) in Indian agriculture. RASHTRIYA KRISHI, 11 (2): 45-49.
Kumar, P. and Namgyal, D. (2020). Retrieved from https://www.earlytimes.in/newsdet.aspx?q=298748
Kumar, R., Kumar, S., Pundir, R. S., Surjit, V. and Rao, C. S. (2022). FPOs in India: Creating en¬abling ecosystem for their sustainability. ICAR-National Academy of Agricultural Research Management, Hyderabad, India.
Kumar Ujjwal, Singh, D, K,, Bhatt, B, P,, Sarkar Bikash, Koley, T. K. and Gupta Santosh (2017). Mar¬ket Led Agricultural Extension-Concept and Practices. Training Manual ICAR Research Complex for Eastern Region, Patna -800 014
MANAGE (2023). Retrieved from (https://www.manage.gov.in/cfa/cfa.asp)
Meena, M. S., Kale, R. B., Singh, S. K. and Gupta, S. (2016). Farmer-to-Farmer Extension Mod¬el: Issues of sustainability & scalability in Indian perspec¬tive. https://www.researchgate.net/publication/313160493_Farmer-to-Farmer_Extension_Model_Issues_of_Sustainability_Scalability_in_Indian_Perspective
MIDH, DAFW, MOA, GOI. 2014. Retrieved from https://midh.gov.in/PDF/midh(English).pdf
Milletodisha.com. (2021). Retrieved from https://milletsodisha.com/uploads/files/wfh/Lesson-from-Odisha-Millets-Mission_Report_final.pdf
Mission Shakti (2022). Annual Activity Report 2021 - 2022 Department of Mission Shakti Government of Odisha. https://missionshakti.odisha.gov.in/sites/default/files/2022-12/Mission%20Shakti%20Activity%20Report%202021-22_compressed.pdf
My Schemes. 2023. https://www.myscheme.gov.in/schemes/atma-scheme
NAAS 2022. Gender and Nutrition based Extension in Agriculture. Policy Paper No. 112, Na¬tional Academy of Agricultural Sciences, New Delhi pp 16
NABARAD (2023). Retrieved from https://www.nabard.org/nabkisan.aspx
NFSM, DAFW, MOA, GOI. 2018. Retrieved from https://www.nfsm.gov.in/Guidelines/NFSM12102018.pdf
NICRA, ICAR. (2012)http://www.nicra-icar.in/nicrarevised/images/Books/brochure.pdf
NIPHM (2023). National Institute of Plant Health Management. Department of Agriculture & Farmers Welfare, Ministry of Agriculture, Government of India. Retrieved fromhttps://niphm.gov.in/general/objective.html
NITI AYOG, GoI.2023. Retrieved from https://niti.gov.in/aspirational-districts-programme
NRAA (2022). Accelerating the Growth of Rainfed Agriculture - Integrated Farmers Livelihood Approach. National Rainfed Area Authority (NRAA) Department of Agriculture and Farmers' Welfare Ministry of Agriculture & Farmers' Welfare. Government of India. Retrieved from https://agricoop.gov.in/Documents/121233187_rapfinaldraft%20(1)_repaired.pdf
Odishaagro.nic.in. 2023. Retrieved from https://odishaagro.nic.in/#/about-us
Osscltd. 2023. Retrieved from https://www.osscltd.in/cms/about-us/introduction
Outlook Planet (2023). Retrieved from https://planet.outlookindia.com/news/odisha-millets-mission-reviving-millets-in-farms-and-plates-news-415168
Padaria, R. N., Ranjith, P. C., Tanwar, R. and Shasani, S. (2022). State of Agricultural Extension reforms in India and the need of convergence. Current Science, 123(3):264.
PIB. 2021. Retrieved from https://pib.gov.in/PressReleaseIframePage.aspx?PRID=1707025

PIB. 2022. Retrieved from https://pib.gov.in/PressReleasePage.aspx?PRID=1823404

PIB. MOA. 2021. Retrieved from https://pib.gov.in/PressReleasePage.aspx?PRID=1783875

PRADAN Annual Report 2021-22. Retrieved from https://www.pradan.net/wp-content/uploads/2022/09/Pradan-Annual-Report___09_09_22-1.pdf

Prasarbharati.gov.in. (2017). Retrieved from https://prasarbharati.gov.in/farm-home/

Rasheed, S. V. (2012). Agricultural extension in India: current status and ways forward. Roundtable on Agricultural Extension in Asia, Beijing, 2013, 15-17.

Ratnakar, R. (2016). Indian Agriculture: Role of extension in the changing agriculture scenario. International Journal of Economic Plants, 3(1):16-22.

Sah, U., Singh, S. K. and Pal, J. K. (2021). Farmer-to-Farmer Extension (F2FE) approach for speedier dissemination of agricultural technologies: A review. The Indian Journal of Agricul¬tural Sciences, 91(10), 1419-1425. https://doi.org/10.56093/ijas.v91i10.117403

Sahoo, A. K., Burman, R. R., Lenin, V., Sarkar, S., Sharma, P. R., Sharma, J. P., and Iquebal, M. A. (2017). Critical analysis of IARI-Post Office Linkage Extension Model and strategies for out-scaling. Indian Journal of Extension Education, 53(4):45-51.

Sahoo, A. K., Meher, S. K., Panda, T. C., Sahu, S., Begum, R. and Barik, N. C. (2020). Critical review on cooperative societies in agricultural development in India. Current Journal of Applied Science and Technology, 39(22), 114-121.

Sahoo, A. K., Sahu, S., Meher, S. K., Begum, R., Panda, T. C., and Barik, N. C. (2021). The role of Krishi Vigyan Kendra (KVK) in strengthening national agricultural research extension system in India. Insights into Economics and Management, 8(9): 43-45.

SAMETI (2023). Retrieved from https://www.sameti.org/what_sameti.php

SAMETI, Jharkhand. (2023). Retrieved from https://www.sameti.org/Guidelines/Policy%20Framework%20for%20Agricultural%20Extension.pdf

Sandhya, V and Nirmala, K. (2018). "Commodity boards in India: Development, Regulation and Current Scenario", International Journal of Emerging Technologies and Innovative Research (www.jetir.org | UGC and ISSN Approved), ISSN:2349-5162, Vol.5, Issue 6, page no. pp409-412, June-2018, Available at : http://www.jetir.org/papers/JETIR1806447.pdf

Saravanan, R., Ashwini, D. and Darshan, N.P., 2020. Agricultural Extension and Advisory Ser¬vices: Serving Farming Community by Agripreneurship Amid COVID-19. Working Paper 4, MANAGE Centre for Agricultural Extension Innovations, Reforms and Agripreneurship, Na¬tional Institute of Agricultural Extension Management (MANAGE), Rajendranagar, Hyderabad.

Sharma, S., Rana, D. S., Jat, M. L., Biswal, S., Khatri-Chhetri, A. and Pathak, H. (2020). A compendium of Technologies, Practices, Services and Policies for Scaling Climate Smart Agriculture in Odisha (India).

Singh & Ranjan. (2017). Retrieved from https://informatics.nic.in/uploads/pdfs/4caab3c2_bihar-state.pdf

Singh, Krishna M., Meena, Mohar Singh and Swanson, Burton E. (2013). Role of State Agricultural Universities and Directorates of Extension Education in agricultural extension in India. Available at SSRN: https://ssrn.com/abstract=2315381 or http://dx.doi.org/10.2139/ssrn.2315381

Vikasrabi.com (2023). Retrieved from https://vikasrabi.com/Home/About

8

Challenges in Agriculture Sector and Role of Financial Institutions in Eastern India

Kamal P. Patnaik

Regional Director for Telengana, Reserve Bank of India, Hyderabad, Telangana, India

1. Introduction

India is a land of great diversities in terms of people, culture, languages and the style of living, among many other varied characteristics in the country. Based on the climatic, geographical, and cultural features, the country is broadly divided into six zones - the Northern, Southern, Eastern, Western zone, Central and the North-Eastern zone. Eastern zone comprises Bihar, Jharkhand, West Bengal, Odisha, Chhattisgarh, Assam and eastern UP (27districts of UP). This document exclusively focuses on the four states namely Bihar, Jharkhand, Odisha and West Bengal in the Eastern Zone with an attempt to look into the bottlenecks inhibiting the rural development of the zone.The financial institutions for agriculture in India was developed with the purpose of financing the needy farmers and providing them the facilities to increase the efficiency of agriculture. As a result, NABARD was formed in 1982. Thus, it was developed as an apex bank to help and support the agriculture section in India.This chapter seeks to evaluate the role of various financial institutions which play a major role in raising the standard of living of nearly one-fourth of India's population in the eastern region of our vast country.

The eastern region lies in the humid-subtropical zone, and experiences hot summers from March to June, the monsoon from July to October and mild winters from November to February during which the whole region receives heavy, sustained rainfall. Major crops cultivated in the region include rice, maize, pulses, tea, jute, rubber, paddy, sunflower, vegetables etc.

The Eastern India occupies about 22% geographical area and supports 34% of the human population of India. The total geographical area of the region amounts to 718.40 lakh ha with the highest in Odisha (155.71 lakh ha). The total population of the region is 4069.20 lakh and with a density of 567 person/sq.km. Bihar state has the highest population of 1040.99 lakh among the states in eastern zone.

The area, population and density of four states of eastern region as compared to eastern zone and India are given in the following Table 1.

Table 1. Area and population (2011).

State	Geographical area (lakh ha)	Population (Lakh)	Density (no/Sq. km)
Bihar	94.16	1040.99	1105
Jharkhand	79.72	329.88	414
Odisha	155.71	419.74	269
West Bengal	88.74	912.76	1029
Eastern India	718.40	4069.20	567
India	3287.26	12108.50	382

Source: Census 2011.

While area and population provide an idea about the demography of the region, a better indication of the productivity of the economic activities may be shown by per capita State Domestic Product.

The net State Domestic Product for the four states along with the national per capita GDPare given in the following Table 2.

Table 2: Per Capita Net State Domestic Product (NSDP) 2020-21 (Current Prices)

Bihar	Rs 43605
Jharkhand	Rs 71071
Odisha	Rs 101501
West Bengal	Rs 121267
India	Rs 146087 (per capita GDP)

Source: Handbook of Statistics on the Indian Economy 2021-22

Bihar has the lowest per capita NSDP. This is probably because of the high denominator effect produced due to the high population of the state. However, other factors such as climatic conditions, availability of infrastructure such as power, irrigation facilities etc. also affect the overall productivity in different components. This is also reflected by the fact that despite higher population and lower area than Odisha, West Bengal has a higher per capita NSDP. This could be attributed to naturally advantageous factors such as above average rainfall received by the state helping it to increase its production of water intensive crops such as rice, sugarcane etc.

The average annual rainfall in the four states for the past period starting from 2016 as compared with India are given in the following Table 3.

Table 3. Annual rainfall (mm).

	2016	2017	2018	2019	2020	2021
Himalayan West Bengal & Sikkim	2624.8	2684.9	2355.3	2472.7	2691.1	2616.7
Gangetic West Bengal	1427.0	1568.6	1225.9	1520.3	1064.6	2112.7
Odisha	1253.5	1344.5	1630.0	1593.9	1140.9	1420.8
Jharkhand	1264.0	1165.8	960.0	1137.8	898.3	1444.8
Bihar	1158.0	1112.0	860.6	1194.7	1272.4	1512.7
India	1083.1	1127.1	1020.8	1284.1	1280.6	1236.1

Source: Handbook of Statistics on the Indian Economy 2021-22.

The above data show evidence with regards to excess rainfall being a contributing factor to higher per capita NSDP. Both divisions of West Bengal received much higher average annual rainfall over the years in comparison to the other states and the national average as well.

Another factor imperative to the growth of agriculture in any region is the financial support provided by the institutions to the sector. Financial Institutions, especially Scheduled Commercial Banks, have been at the forefront of lending to the Agriculture Sector in India. To compare the level of credit reach to the agriculture sector in these states, we look at the per capita credit deployed in the agriculture sector.

The projected adult population by 2021, credit to agricultureby SCB (crores) & per capita credit to agriculture (Rs) in four states as compared to India are given in the following Table 4.

Table 4. Per Capita credit to Agriculture.

State	**Projected adult population 2021[@] (Lakhs)**	**Credit to Agriculture by SCBs[#] (Crores)**	**Per Capita credit to agriculture (Rs)**
Bihar	734.49	49605	6753.67
Odisha	313.69	23978	7643.85
Jharkhand	248.63	10390	4178.90
West Bengal	726.51	46372	6382.84
India	9389.59	1518112	16168.03

Source: @ Populations Projections report, Ministry of Health & Family Welfare # Handbook of statistics on Indian States 2021-22

The data shows that Odisha has a relatively higher per capita credit deployed in Agriculture. However, all the states fall significantly short when compared to the national average in this regard. Another interesting factor to be noted here is that despite having comparable levels of per capita credit, West Bengal had a relatively higher per capita NSDP (Table 2). This again provides substance to the fact that other factors such as climatic zones, support from Governments, other agencies such as NABARD, NGOs etc. are also crucial to the growth of output and subsequently productivity in different sectors.

The degraded land, net area sown, cropping intensity and irrigated area are important agriculture profile. The degraded land (%), net area sown (lakh ha), cropping intensity (%) and irrigated area (%) of four districts as compared with eastern zone and India during 2018-19are given in following Table 5.

Table 5. Degraded land (%), Net area sown (lakh ha), Cropping intensity (%) and Irrigated area (%) during 2018-19

State	Degraded land (%)	Net area sown (lakh ha)	Cropping intensity (%)	Irrigated area (%)
Bihar	18	56.65	139	61.1
Jharkhand	54	15.36	116	12.0
Odisha	32	56.24	160	37.0
West Bengal	25	54.85	150	57.0
Eastern India	-	312.17	150	-
India	47	1395.61	139	44.0

Source: (i) Agricultural Statistics at a Glance,2021, GOI (ii) Fertilizer and agriculture Statistics, Northern Region, FAI, New Delhi,2018-19.

The degraded land of Jharkhand is 54 %, being highest among four states of eastern zone and even higher than all India average of 47%. The cropping intensity of Odisha is 160 %, being highest among four states of eastern India and even higher than all India average of 139%. Besides the cropping intensity of West Bengal (150%) is higher than all India average. Bihar state has highest irrigated area of 61.1% among four states of eastern India followed by 57% in West Bengal as compared to 44% irrigated area of all India average.

Food grain production is the most important activity in eastern India, which provides income and employment to a larger section of the population. Area (lakh ha) & production (lakh t) of food grains, rice of four states of eastern India as compared with eastern India and allIndia during 2016-17 have been given in following Table 6.

Table 6: Area (lakh ha) & production (lakh t) of food grains, rice of four states of eastern India as compared with eastern India and all India during 2016-17.

State	Food Grains		Rice	
	Area	Prod	Area	Prod
Bihar	66.80	165.30	33.40	82.40
Jharkhand	30.50	56.60	16.70	38.40
Odisha	63.70	116.80	39.60	97.90
West Bengal	62.60	171.40	54.00	153.00
Eastern India	374.40	867.10	237.60	579.20
India	1292.30	2751.10	439.90	1097.00

Source: (i) Agriculture Statistics at a Glance, 2018, GOI, Odisha Agriculture Statistics 2016-17 (ii) Hand Book of Statistics on Indian states, RBI, Nov 24,2021, (iii) Fertilizer and Agriculture Statistics, Northern region, FAI, New Delhi, 2018-19.

India produced 2751.1 lakh t of food grains from an area of 1292.3 lakh ha during 2016-17. Eastern India produced 867.1 lakh t of food grains, 31.51% of India's production during 2016-17. The production (lakh t) of food grains in different regions of eastern India are Bihar (165.3) from an area of 66.8 lakhha, West Bengal (171.4) from an area of 62.6 lakh ha, Odisha(116.8) from an area of 63.7 lakh ha, and Jharkhand (56.6) from an area of 30.5 lakh ha As against food grains productivity of 2129 kg/ha in India during 2016-17, the productivity (kg/ha) of the major states of West Bengal (2738), Bihar (2474), Eastern UP (2907), Jharkhand (1855), and Odisha (1833).

Rice contributed about 67% of the total food grains production in eastern India as compared to 40% in India during 2016-17. Rice is the principal crop in eastern India, which occupies 237.6 lakh ha accounting for 54% of the total rice-growing areas of the country. Out of total rice production of 1097.0 lakh t in the country from an area of 439.9 lakh ha, the production of the Eastern India region is 579.2 lakh t from an area of 237.6 lakh ha, which is about 53% of India's production during 2016-17. The production (lakh t) of rice in the regions are West Bengal (153.0 lakh t) from an area of 54.0 lakh ha, Odisha (97.9 lakh t) from an area of 39.6 lakh ha, Bihar (82.4 lakh t) from an area of 33.4 lakh ha and Jharkhand (38.4 lakh t) from an area of 16.7 lakh ha. The productivity (kg/ha) of rice in the major states of Eastern India during 2016-17 are West Bengal (2833), Odisha (2472), Bihar (2467) and Jharkhand (2299) as against National Productivity of 2494 kg/ha.

2. Support from thefinancial Institutions to the Agricultural Sector

With the Government pushing towards increasing institutional credit to the agricultural sector, several financial institutions have been involved in increasing their reach to the agricultural sector through different modes and have been engaged in providing timely credit to this sector. The following are the major financial institutes that can play a major role in deepening the reach of the formal finance to the agricultural sector through various initiatives.

i) **NABARD:** This is the apex banking institution set up to provide finance for Agriculture and Rural development and fostering rural prosperity. NABARD promotes sustainable and equitable agriculture and rural development through participative financial and non-financial interventions, innovations, technology and institutional development for securing prosperity. NABARD makes its interventions through various direct and indirect schemes of GoI and also through Rural Infrastructure Development Fund (RIDF) which is used for financing various developmental projects in the states. The eligible activities for RIDF are classified under three broad categories i.e. Agriculture and related sector, Social sector and Rural connectivity. State Governments/ Union Territories, State Owned Corporations / State Govt. Undertakings, State Govt. Sponsored / Supported Organisations etc. are some of the eligible institutions who can avail support from RIDF for various projects. NABARD also supervises cooperative banks and Regional Rural Banks along with providing them refinance support.Among many developmental agencies, NABARD has a particularly crucial role to play in development of rural India. The State Focus Paper (SFP) of NABARD lays out projections for realizable potential in different sectors for the state. It also outlines the various directed interventions required to be made by the different stakeholders including the State/UT departments.

Table 7: NABARD's projected potential in agriculture& credit deployed by SCBs

State	Projected Potential@	Credit to Agriculture by SCBs#
Bihar	87874	49605
Jharkhand	12023	23978
Odisha	52050	10390
West Bengal	97261	46372

Source: @ NABARD's State Focus Paper 2023 for respective States

Handbook of Statistics on Indian States 2021-22

A brief comparison of the projected potential in agriculture and credit deployed to this sector by Scheduled Commercial Banks (SCBs) indicates that there is still some gap between the identified potential and deployed credit. There is, therefore, sufficient scope for bridging this gap through guided efforts of the stakeholders aimed at raising the agricultural productivity. Having the necessary expertise and skillset in developing solutions for rural India, other agencies need to work in coordination with NABARD to realize the vision of a prosperous rural India.

ii) **Commercial Banks:** These Banks have been mandated by the Reserve Bank of India to lend 40% of their Adjusted Net Bank Credit (ANBC) or Credit Equivalent Amount of Off-Balance Sheet Exposures (CEOBE) to Priority Sector which includes lending to Agriculture (18%), Micro Enterprises (7.5%), Weaker Sections (12%) etc. Although the banks do lend credit to the agriculture sector, the presence of the banks needs an uplift to increase the per capita credit deployed to productive levels in the Eastern Zone.As of December 2022, there are a total of 11,753 rural branches of the commercial banks (including Small Finance Banks, Payment Banks, RRBs etc.) in these four states with the number of commercial banks being 29, 24, 33 and 34 in Bihar, Jharkhand, Odisha and West Bengal respectively.

iii) **Regional Rural Banks (RRBs):** The RRBs combine the characteristics of a cooperative in terms of the familiarity of the rural problems and a commercial bank in terms of its professionalism and ability to mobilise financial resources. The directed lending of RRBs to the agricultural sector over the years has ensured that credit flow to various segments under this sector has continued unhindered. Unlike commercial banks, which have a target of 40% for lending to Priority Sectors, the RRBs must lend 75% of their Adjusted Net Bank Credit (ANBC) or Credit Equivalent Amount of Off-Balance Sheet Exposures (CEOBE) to the Priority Sector. There are a total of 8 RRBs in these four states attributing to 4497 rural branches as on December 2022.

iv) **Cooperative Banks:** These Banks are established with the motto of 'no-profit-no-loss' and thus, do not actively seek profitable ventures. The cooperative banks play a very active role in extending agricultural credit to the farmers and have been the preferred choice of farmers in many areas in the country. They are especially critical in rural areas with scarce banking facilities where they act as the only source of financial support to the farmers. These institutions play a critical role in last-mile credit delivery and in extending financial services across the length and breadth of the country through their geographic and demographic outreach. These banks have also managed to maintain a personal rapport with their customers. Cooperative banks include Primary Agricultural Credit Societies (PACS), State Cooperative Banks, District Central Cooperative Banks etc.

v) **Small Finance Banks:** These Banks are the financial institutions that provide financial services to the unserved and unbanked region of the country. They are registered as a public limited company under the Companies Act, 2013. They are required to extend 75% of their Adjusted Net Bank Credit (ANBC) to the sectors eligible for classification as priority sector lending by the Reserve Bank of India. A total of 8 Small Finance Banks have their presence in this zone with a total of 300 rural branches. With a spurt in growth of these banks in the recent times, this number is expected to grow manifold in the coming future.

vi) **Micro Finance Institutions (MFIs):** MFIs provide loans (usually up to Rs 50000) to the poor farmers without collateral along with flexible EMI options. However, the interest rates are relatively higher than banks as MFI can't accept deposits. They arrange funds via banks/NBFC/ All India Financial Institutions (AIFI) & keep their profit margin by lending these funds further to customers. As far as the presence of MFIs in these four states is concerned, as per RBI data, West Bengal has 10 NBFC-MFIs registered in the state while Bihar and Odisha have one and two MFIs registered with RBI. Jharkhand does not have any registered MFI. However, there may be number of other MFIs not registered with RBI that may be catering to the rural population. As per the Directory of MFIs in India report by SIDBI, in 2014 there were 11, 8, 29 and 45 MFIs in Bihar, Jharkhand, Odisha and West Bengal respectively. These MFIs were in different legal forms such as NBFC, Society, Trust, Section 25 company, Cooperative etc.

To get an indicative picture of the credit disbursed to the agricultural sector in the four states of the Eastern Region, following table is presented with the data on credit to agricultural sector by the Scheduled Commercial Banks which include PSBs, Private Sector Banks, RRBs, Small Finance Banks etc.

Table 8: Credit to Agriculture by Scheduled Commercial Banks (2022)

State	**Credit Extended (crores)**	**No. of branches**	**Credit per branch (crores)**
Bihar	54380	7640	7.11
Jharkhand	12221	3206	3.81
Odhisa	31015	5367	5.77
West Bengal	47955	9616	4.98
Eastern India	145571	25829	5.63
India	1703315	154758	11.00

Source: RBI's handbook of statistics on Indian states 2021-22

The above table shows that Jharkhand and Odisha have relatively low banking presence indicated by the low number of branches. However, notwithstanding the low number of branches, the credit extended per branch is also low in these two states with Bihar recording the highest credit per branch among the four states. All the four states combined have an average credit per branch of 5.63 crore while the national average stands as 11 crore per branch. This reflects that the credit penetration to the agricultural sector by the SCBs in these states needs to be enhanced. Therefore, the total credit extended as well as the bank branches needs improvement in these states.

It won't be incorrect to say that among all the stakeholders involved in contributing to the growth and prosperity of the sector, the Government and its affiliated agencies have a very important role to play. No initiative can take form without the support of these agencies. The Government, on its part, has been active in providing direct and indirect support to the sector through various reforms and schemes such as agricultural initiatives through five-year plans, Land reforms, encouraging innovation in irrigation methods, Scientific farming, Marketing support for Agriculture etc. Schemes such as Kisan Credit Card, Pradhan Mantri Fasal Bima Yojna, Interest Subvention Scheme etc. have also been instrumental in improving the economic status of the farmers. Although the income of the farmers has risen from what they were a few years ago, it is imperative that the stakeholders continue their efforts for leveraging technological solutions for improving the livelihoods of the marginalised strata of the society.

3. Structural Bottlenecks in Rural Agricultural Development

While the institutional credit to the agricultural sector has increased manifold over the years through directed interventions by various stakeholders such as the Government of India, state Governments, RBI, NABARD etc., there remain many challenges to be overcome to ensure steady access of timely and adequate credit to this sector.

a) **Inadequate banking penetration and lack of sufficient institutional credit:** Banking presence, through either brick-and-mortar branches, or other modes such as Business Correspondents, Customer Service Points etc. is essential to ensure availability of credit in the rural pockets of the country. Although the number of bank branches have increased substantially over the last few years, banks need to continuously increase their presence in the rural areas of the country to link the agriculture-dependent population to the formal finance system.

Table 9: Bank branches per lakh population in Eastern India

State	Projected Population[@] (2022) in crores	Commercial Bank branches (2022)[#]	Branches per lakh population
Bihar	12.49	7612	6.09
Jharkhand	3.89	3184	8.18
Odisha	4.59	5375	9.53
West Bengal	9.86	9424	9.55
Eastern India	30.83	25595	8.30
India	137.58	153639	11.16

Source: @National Commission on Population's (Ministry of Health & Family Welfare), Report on Population Projections

DBIE, RBI (banks include SFBs, Payment Banks, LABs, Foreign Banks among others)

b) **Low agricultural productivity:** Though the country has come a long way from famines and food shortages occurring decades ago, increased production levels have also been matched by the increased population levels in the country. Food shortage doesn't appear a significant concern now considering the production capabilities of India in food grains butgiven that the proportion of cultivable lands are largely expected to remain static, the margins of increased production would come to a grinding halt in the absence of persistent measuresaimed at improving productivity. This requires multiple inputs, agricultural credit being one of them. Along with credit, investment needs to be made in developing new varieties of crops, seeds, cropping methods and long-term capital investment needs to be encouraged in agricultural infrastructure. Projects in best new practices in agriculture should be encouraged and the last mile farmer should be adequately trained wherever necessary.

Table 10: Yield (kg/ha) for major crops in Eastern India (2016-17)

Crop	Bihar	Jharkhand	Odisha	West Bengal	India
Rice	2467	2299	2472	2833	2494
Wheat	2427	2011	1333	2682	3199
Bajra	1135	667	614	293	1305
Maize	3732	1923	2827	4367	2689
Pulses	901	1000	502	962	785
Gram	1120	1174	768	1013	974
Oilseeds	1155	722	889	1139	1192
Total Foodgrains	2474	1855	1833	2738	2129

Source: Agricultural Statistics at a Glance, Ministry of Agriculture and Farmers Welfare, Government of India

The above table presents data on yields of major crops for the Eastern states and the country. The data indicates a mixed picture. Expectantly, yield for Rice is higher than the national figure in West Bengal predominantly due to higher rainfall received by the state and it is lower in the remaining three states. Further, Wheat, Bajra, and Oilseeds have lower than national yields in these states. Jharkhand and Odisha have relatively lower yields in most of the crops. Nevertheless, the need for improving the yield is very relevant at this point of time when there is a consistent increase in population of the country. Add to this the fact that a large chunk of the population is dependent on agriculture for their livelihood, it is imminent that measures for strengthening this sector are taken timely.

As noted above, variety of other factors also contribute to the dynamics of the productivity in the agriculture. Some of the states such as West Bengal are naturally endowed with favorable climatic conditions for crops such as rice, cotton and other water intensive crops. In comparison, states such as Bihar, Punjab, UP etc. house a high number of population but receive less than average rainfall. Given that most of the population is engaged in agriculture, this forces constraints on the type of crops farmers can cultivate in these regions. This may also push farmers residing in such regions to move out of state in search of better opportunities in Agriculture or other sectors. As per Census 2011 data, UP and Bihar had a disproportionately high number of migrants. The two states together accounted for 37% of the total inter-state migration. Even though most of the migrants eventually settle for laborious jobs in factories, industries and other unorganized segments, some population may still search for livelihood in agriculture in other better productive states. This does present an opportunity for the various State and Central agencies to encourage innovation led by technology in the agriculture sector so that new methodologies can be brought about which could raise agricultural productivity in these relatively dry states.

c) **Infrastructural hurdles:** Significant development has been made in creation of essential infrastructure in the country in the form of roads, rail network, electricity etc. However, the agricultural growth also suffers from impediments in the form of infrastructural bottlenecks such as lack of adequate cold chain storage capacity, processing facilities, marketing support, inadequate irrigational facilities and high dependence on monsoon etc. Merely providing basic facilities such as connectivity and power availability may not be enough to derive the maximum of potential available in agricultural output. More efforts need to be invested in providing supplemental facilities to raise the levels of agricultural output.

d) **Fragmented Landholdings:** Over the years, the number of farm holdings in India has increased but the area under farming has come down. Marginal and small holdings together constitute 86 per cent of total holdings in India. Such fragmentated land occupancy structure makes it almost impossible for farmers to viably make certain capital investments such as in tube wells, drip irrigation, storage or bulk inputs. Land consolidation is thus important to drive higher efficiency. Unfortunately, the land market in India is thin and various constraints such as poor quality of land records, complex administrative procedures, restrictions on transferability of land rights, leasing, sub-leasing and rental arrangements hinder the development of land market. There is a need for consolidation of landholdings through land market reforms to increase farm productivity in the country.As per NABARD's study on the Agrarian Structure and Transformation of the Institutional Framework of Agriculture Sector, the average operational land holdings of the farmers in the Eastern India are as displayed in the table below. The data indicates that barring Jharkhand, all remaining three states have average holdings below that of National Average of 1.08 hectares.

Table 11: Average size of operational land holdings

State	Average size of holdings (hectare)
Bihar	0.39
Jharkhand	1.10
Odisha	0.95
West Bengal	0.76
India	1.08

Source: A Study of the Agrarian Structure and Transformation of the Institutional Framework of Agriculture Sector Using Data from Agricultural Censuses (NABARD).

Further, there is also a need for pushing the pace of digitalisation of land records in the country which is taking place under the ambit of Digital India Land Records Modernization Programme (DILRMP – MIS 2.0). Digitalisation of records is important for deepening the reach of institutional credit to the agriculture sector. All the states/UTs have been advised by the Centre to undertake digitalisation of records under this programme. The credit outreach to this sector is expected to improve along with other benefits such as creation of charge digitally, once the exercise is completed.

e) **Volatility in Food Prices:** As agricultural production in India is still heavily dependent on rainfall and its spatial distribution, adverse climatic conditions like draught, flood and unseasonal rains tend to disrupt both aggregate supply and supply chains, imparting large volatility to food inflation trajectory. The contribution of food, and particularly, vegetables, to volatility in headline inflation is significantly higher than non-food items, reflecting the perishable nature of the crop, short crop cycles, lack of adequate storage and poor pre- and post-harvest practices. The role of supply management in terms of minimising post-harvest losses, scaling up storage infrastructure, development of food processing industry and upgrading food safety standards assumes importance in this context. Public and private sector investment can play a critical role to build up the agricultural infrastructure required for efficient post-harvest practices.

f) **Low Public Investment:** Capital formation holds the key to agricultural growth and sustainable development. Capital formation in Indian agriculture as a per cent of agriculture GVA, however, has moderated in the recent period. The share of public sector is much lower than private sector in capital formation in agriculture. While public investment in agriculture primarily constitutes infrastructure development, irrigation, research and extension services, private investment - dominated by households - is made either for augmenting productivity of natural resources or for supplementing income sources of farmers from allied activities. Public investment has largely hovered around 1.7- 3.5 per cent of agriculture GVA during the last two decades. Although, leading other countries such as Brazil and Russia, India lags much behind China, which is more representative of the challenges our country faces such as growing population.

Table 12: Gross Capital Formation in Agriculture, Forestry and Fisheries

Country	Gross Fixed Capital formation in Agriculture, Forestry and Fisheries, 2021 (in million $)
Brazil	4238.69
Russia	11705.27
India	65383.15
China	194519.74
South Africa	1713.42
Pakistan	8256.23

Source: Food and Agriculture Organization, United Nations.

4. The Way Forward

Indian agriculture scaled new heights with recordproduction of various foodgrains, commercial andhorticultural crops, exhibiting resilience and ensuringfood security for the country. The UN's Food and Agricultural Organisation (FAO) published a report titled World Food and Agriculture Statistical Yearbook in 2022 which shows the production of primary crops and commodities for some of the major agricultural countries as below:

Table 13: Production of Primary Crops & Commodities, 2020

Country	Production of Primary Crops & Commodities (thousand tonnes)
Brazil	1098025
Russia	228230
India	1113476
China	1816320
South Africa	51576
US	710680
Germany	93832

Source: World Food and Agriculture Statistical Yearbook, FAO, UN.

Holistic Policy Approach: The sector, however, confronted various challenges, mitigation of which requires a holistic policy approach. For instance, crop productivity in India is lower than otheradvanced and emerging market economies due tovarious factors, viz., fragmented landholdings, lowerfarm mechanisation and lower public and privateinvestment in agriculture. As per Food and Agriculture Organization, UN, in 2020, the yield for Wheat in India was 3.43 (Tonnes per Hectare), 4.14 in South Africa and 5.74 in China. Similarly, the yield in Rice was 3.96, 2.82 and 7.04 in India, South Africa and China respectively.

Boost the Production of Fibre Crops, Tea, Coffee & Rubber: Further, current over-production of crops like rice, wheat and sugarcane, has led to rapid depletion of ground water table, soil-degradation and massive air pollution raisingquestions aboutenvironmental sustainability ofcurrent agricultural practices in India. Considering the agro-climatic zone of the Eastern India, the region is better placed to invest in production of crops such as Tea, Coffee, Rubber, Jute, Cotton etc.

Cluster- Based Approach in Agriculture Production: An innovative way to raise the production and subsequently the rural income levels can be through setting up of clusters for cash crops. These clusters can not only help provide employment to the rural households but will also help provide a smooth transition from the traditional low yield crops to the high-income generating crops. Special efforts need to be directed to set up these clusters in the rural areas.

Formation of Farmer Producer Organizations (FPOs): To address the viability concerns for capital infrastructure due to fragmented landholding, the aggregation of produce model in the form of Farmer Producer Organisations (FPOs) can be adopted. This model will help farmers reap the dividends of economies of scale and also provide an optimum return to the on their produce by minimizing the costs through collectivization of efforts.

Promotion of Milles: Millets are important staple cereal crops for millions of smallholder dryland farmers across Asia and sub-Saharan Africa. They are also called nutri-cereals or dryland cereals, and include sorghum (jowar), pearl millet (bajra), finger millet (ragi), foxtail millet , proso millet, barnyard millet and kodo millet etc and offer high nutritional benefits. Millet crops requires less water and are more resilient to climate vulnerability. The United Nations General Assembly adopted an India-sponsored resolution to mark 2023 as the international year of millets. It is pertinent to note that India is the largest producer of millets. Millets have long been the staple food of the tribal population in India. In 2020, India accounted for around 41% of the total production of millets in the world. The major millets producing states in India are Rajasthan, Uttar Pradesh, Haryana, Gujarat, Madhya Pradesh, Maharashtra, Karnataka, Tamil Nadu, Andhra Pradesh and Telangana. Three states namely Rajasthan, Uttar Pradesh and Haryana account for more than 81 per cent share in total millet products. Millets have been accorded a prominent position in this year's budget. The Government's dedication to spreading the use of millets was made clear in the Union Budget 2023 when Finance Minister Nirmala Sitharaman declared funding for turning the millet institution in Hyderabad into a centre of excellence. Millet crops can be grown successfully in eastern Indian states. The Government of Odisha launched the special programme for promotion of millets in tribal areas known as Odisha Millet Mission (OMM) in 2017 with aim to Revive Millets on Farms and Plates and simultaneously focus on production, processing, consumption, marketing and inclusion of millets in Government Schemes. The Government of India has asked all states to adopt Odisha Millets Mission model for promotion of millets. Odisha government has procured over six lakh quintals of ragi or finger millet under Odisha Millet Mission (OMM) from over sixty thousand farmers in the Kharif 2022-23. The crop was acquired at a minimum support price (MSP) of Rs 3,578 per quintal from 143 blocks in 19 districts. The Odisha government is planning to establish a benchmark price equivalent to the Minimum Support Price (MSP) for non-ragi millets, such as foxtail millet and little millet, which are not currently procured under the MSP. Odisha has emerged as one of the fore runners of millets, ensuring the participation of women Self-Help Groups (SHGs) under Mission Shakti(shelf help mission) in the millet value chain where they are playing a leading role in processing, value addition, and marketing.

Joint effort of Government, Bank and Stakeholders: The power of collaboration between the government, bank and stakeholders needs to be systematically harnessed for realizing the vision of a new resurgent India. Availability of infrastructure will not yield the desired results if institutional finance is not available on a timely basis. Mobile modes of banking such as Business Correspondents and Customer Service Points need to be encouraged, especially in the far-flung rural areas where setting up of brick and mortar branches is often not feasible. Financial institutions, thus, need to reset their focus towards agriculture well beyond the parameters as mandated by RBI under the Priority Sector norms.

Focus on Sustainable Agriculture: Despite surplus production in many of the commodities, food inflation and volatility in prices continue to remain high causing inconvenience to consumers and low and fluctuating income for farmers on a real-time basis. Addressing these challenges would require a second green revolution focussed on the agriculture-water-energy nexus, making agriculture more climateresistant and environmentally sustainable. The useof biotechnology and breeding will be importantin developing eco-friendly, disease-resistant, climate-resilient, more nutritious and diversifiedcrop varieties.

Wider Use of Digital Technology: Wider use of digital technology andextension services would be helpful in informationsharing and generating awareness among the farmers.A strong push also needs to be given to the efforts for Financial Inclusion in the country, particularly in the rural pockets. To connect the informal sector to the formal channels of credit, a basic bouquet of financial services needs to be provided to the rural households. The Government and Reserve Bank have been proactive in their efforts in this regard by urging banks and other agencies to increase their outreach through various means in the yet to be connected areas. Eventually, the true potential of agriculture in the region can only be harnessed through collaborative efforts of all the stakeholders. However, there's a greater need today to look at the sector through the lens of modern methodologies combining a mix of technology and innovation on a wider scale. SLBC Convenor banks in the Eastern India, i.e., SBI (Bihar), Bank of India (Jharkhand), UCO bank (Odisha) and PNB (West Bengal), have a major role to play and need to be actively involved in collaborative efforts with other agencies such as NABARD, state agencies etc. in realizing the true potential of Agriculture in this part of the country.

References

Agricultural Statistics at a glance - https://eands.dacnet.nic.in

Budget 2023 article dated February 01, 2023 – https://thehindu.com

Food and Agricultural data – https://fao.org

Government of India Census 2011 – https://censusindia.gov.in

Government of India Digitalization of land records programme – https://dilrmp.gov.in

Handbook of Statistics on Indian Economy 2021-22 – https://rbi.org.in, https://dbie.rbi.org.in

Handbook of Statistics on Indian States 2021-22 - https://rbi.org.in, https://dbie.rbi.org.in

International Year of Millets 2023 (Ministry of Agriculture and Farmer's Welfare's Press Release dated January 01, 2023) – https://pib.gov.in

NABARD's State Focus Paper – https://nabard.org

Report on Population Projections – https://main.mohfw.gov.in

Study of the Agrarian Structure and Transformation of the Institutional Framework of Agriculture Sector Using Data from Agricultural Censuses – https://nabard.org

The contribution of Shri Kirty Bhushan, Manager, Financial Inclusion & Development Department of RBI is gratefully acknowledged.

World Food and Agriculture Statistical Yearbook – https://fao.org

9

National Agroforestry Policy 2014 in India

A.K. Sahoo, V.P. Singh* and D. Nayak**

Former Professor and In Charge, Department of Natural Resource Management College of Forestry, Odisha University of Agriculture & Technology, Bhubaneswar Odisha

**Former Regional Representative, ICRAF South Asia, New Delhi*

***The Food and Agriculture Organization (FAO), New Delhi*

1. Introduction

The National Agroforestry Policy of India is a comprehensive policy framework designed to improve agricultural livelihoods by maximizing agricultural productivity including animal productivity along with tree cover and mitigating climate change. The Government of India launched the policy in February 2014 during the World Congress on Agroforestry, held in Delhi. The focus of the policy is to address Agroforestry, which is an integrated system of practicing agriculture, livestock farming and forest activities on the same unit of land, storing high carbon, there by mitigating climate change. Agroforestry systems have the potential to be more beneficial than traditional agriculture and forest production practices. They may be able to give enhanced productivity, as well as social, economic, and environmental advantages, as well as a broader variety of ecological products and services. It's important to remember that these advantages are contingent on appropriate farm management. In 2014 India became the first nation in the world to adopt an inter-sectoral policy on agroforestry, which seeks to identify the bottlenecks in expansion on agroforestry in the country and pathways to remove the constraints in a systematic manner. The national policy puts emphasis on bridging forestry, agriculture, water and the environment, which all deal with the way people interact with land-based resources in the country. Rather than defining agroforestry as a separate set of activities worthy of policy support, the process was geared towards ensuring that other policies do not undermine agroforestry as a solution to multiple national goals. The policy is not only seen as crucial to India's ambitious goal of achieving 33 percent tree cover, but also to providing many of the other benefits such as increasing food and nutrition, and supplying fodder, fuel wood and timber for India's growing population. Agroforestry can become an important tool to build resilience of farmers and rural people against threats of climate change and natural calamities. This can also help in greening the rural employment and rural development opportunities by providing agroforestry tree produce based economic opportunities.

2. Need for Agroforestry Policy in India

Absence of a dedicated and focused national policy on agroforestry and a suitable institutional mechanism in the country: Major policy initiatives, including the National Forest Policy 1988, the National Agriculture Policy 2000, Planning Commission Task Force on Greening India 2001, National Bamboo Mission 2002, National Policy on Farmers, 2007 and Green India Mission 2010, emphasize the role of agroforestry for efficient nutrient cycling, organic matter addition for sustainable agriculture and for improving vegetation cover. However, agroforestry has not gained the desired importance as a resource development tool due to various factors. Some of these factors include: restrictive legal provisions for harvesting & transportation of trees planted on farmlands and use of non-timber produce, near non-existent extension mechanisms, lack of institutional support mechanisms, lack of quality planting materials, inadequate research on agroforestry models suitable across various ecological regions of the country, inadequate marketing infrastructure and price discovery mechanisms, lack of post- harvest processing technologies, etc. This is also due to the fact that the mandate of agroforestry falls through the cracks in various ministries, departments, agencies, state governments, etc. The value and position of agroforestry is ambiguous and undervalued, and despite of its numerous benefits, it is only sporadically mentioned at the national level, because of the lack of appropriate public policy support. While there are many schemes dealing with tree planting / agroforestry, there is an absence of a dedicated and focused policy, and lack of an institutional mechanism for coordination and convergence among the schemes/ ministries to pursue agroforestry in a systematic manner.

Lack of an integrated farming systems approach: Farming enterprise of small farmers needs to be understood and developed as a portfolio of activities rather than as "fixed one type of cropping system". Development along this direction requires a convergent programme which integrates trees, crops, water, livestock and other livelihood initiatives. This perspective of integration seems to be missing in the national agroforestry initiatives in whatever form it may currently be. In fact the key mantra of the success of the agro-horticulture programme of BAIF, NABARD, poplar based commercial scale (though small holder based), agro-timber systems in north-western parts of the country and other successful initiatives is their ability to integrate various livelihood aspects with the tree planting in the farm. Survival of trees is one of the most challenging tasks in the establishment phase of the trees, and without addressing the issue of water this does not seem to be possible. The enthusiasm of farmers depleted substantially with the higher mortality rate as experienced from various programmes in the past.

Restrictive regulatory regime: There are restrictions imposed by the state governments on harvesting and transportation of agroforestry produce, especially those species which are found growing in the nearby forests. These restrictions were basically designed to prevent pilferage from government forests. However, the rationale for such restriction is not very convincing as the species grown in the forest are to be best grown in the nearby private farms because of their suitability to that agro-climatic condition. Obtaining permits for harvesting and transportation are cumbersome, costly& frustrating, and hence, discourage farmers from undertaking tree planting on farm lands. Multiple agencies, including the State Revenue Department are involved in issuing these permits. Similarly, tax is imposed at various stages of the processing by multiple agencies. These restrictions also negatively impact the in-situ, or on-farm primary processing, jeopardize local employment in these operations and increase transport cost because of the transportation of the entire bulk raw material to the processing centers. As a result, the 0domestic agroforestry produce (raw materials and finished goods) is increasingly losing grounds against the imported materials, which are cheaper and of better quality. India, having all the natural advantages, should be able to develop agroforestry as a major sector for income and employment generation.

Inadequate attempts at liberalization of restrictive regulations: There are sporadic examples of States taking steps for liberalization of above restrictions, such as, exempting agroforestry species from the harvesting and transit, but this has not been uniformly done by all the States. Also the extent of liberalization is not widely known to the farmers and thus, their problem continues. It is also learnt that farmers do not take interest in tree planting on the farm land fearing that too many trees on farm may lead to change in their land-use. Clearly such apprehensions have no basis; however this does emphasize the lack of awareness that persists on the ground. The Arun Kumar Bansal Committee, appointed by the Ministry of Environment and Forests in 2011 in its report has also identified the regulatory bottlenecks, impeding the growth of the agroforestry, which need to be acted upon.

Insufficient research, extension and capacity building: Research results on agroforestry, available in the public and private domain do not regularly reach the farmers due to lack of a dedicated extension system. There is a serious lack of institutional mechanisms at all levels to promote agroforestry. The efforts to dovetail agroforestry programmes to any other established programmes which have strong institutional mechanism up to the implementation level, such as the Integrated Watershed Management Programme are non-existent. Also, there is not enough research on the agroforestry models suitable for the diverse agro-climatic regions; for the indigenous and multi-purpose species (viz. Prosopis cineraria) or on domestication of species, resulting in over emphasis on few species (poplar, eucalyptus, Kadam, etc.) and their limited varieties in certain pockets of the country. It is also important to note that India lacks processing technologies for fast growing timber species.

Dearth of quality planting material: Planting material such as seeds, seedlings, clones, hybrids, improved varieties, etc. are generally of mixed quality and not available commonly, particularly in the resource poor regions. It is estimated that only about 10% of planting material is of high quality, the rest without any guarantee for quality standards. This issue mainly relates to the production, handling, distribution and planting & supervision of high quality planting material.

Institutional finance and insurance coverage: Institutional finance in agroforestry has not been at par with its potential due to the lack of awareness of technical and economic data on different agroforestry models, and the techno-economic parameters required by financial institutions (FI) to evaluate finance needs and viability of the projects. Similarly, little is done in developing and popularizing insurance products for agroforestry ventures. Lack of awareness, unavailability of products suitable to growers, high cost of premium and unclear procedure of claim settlements are reported to be the factors responsible for this poor state of affairs.

Weak market access for agroforestry produce: The marketing infrastructure (market yard, etc.), including "price discovery" mechanisms for agroforestry produce in general are unavailable in the country except in few states which have either developed exclusive marketing infrastructure for agroforestry produce or have dovetailed with the regulated agriculture commodity marketing systems. As a result, it is largely a buyer's market and the middlemen get the major share in profit.

Industry operations at a sub-optimal level: The Wood Based Industries (WBI) have played an important role in the promotion of agroforestry and economy in Punjab, Haryana, and in parts of U.P. and Uttarakhand. However, the regulations governing this industry have become stringent. The procedure for setting up new units or fulfilling of compliance by existing units is cumbersome and time consuming, not very encouraging to instil confidence in industries. The restrictions on primary processing at production sites after harvesting, leads to higher cost for transporting entire stock to the factory. This also results in lower supply of raw material, forcing the WBI to operate at sub-optimal level. The role of industries in promotion of agroforestry cannot be ignored and therefore, issues preventing growth need to be addressed urgently. Nearly $ 7-8 billion worth of wood-based products are being imported annually. The low import tariff for raw materials and finished goods, cumbersome procedures for sourcing raw materials domestically are some of the major reasons for the slow or negative growth of the WBI in India. Therefore, the agroforestry policy should facilitate that products are developed at competitive prices within India for generating local employment and reducing burden on imports.

It could be summarized that although farmers are interested to expand agroforestry, as the evidence on adoption shows, there are many missed opportunities for agroforestry to benefit farmer income and the environment due to neglect /oversight of the agencies that are supposed and expected to adequately promote and support it.

3. Goal

The major policy goals are:

- Setting up a National Agroforestry Mission or an Agroforestry Board to implement the National Policy by bringing coordination, convergence and synergy among various elements of agroforestry scattered in various existing, missions, programmes, schemes and agencies pertaining to agriculture, environment, forestry, and rural development sectors of the Government.
- Improving the productivity; employment, income and livelihood opportunities of rural households, especially of the smallholder farmers through agroforestry.
- Meeting the ever increasing demand of timber, food, fuel, fodder, fertilizer, fibre and other agroforestry products; conserving the natural resources and forest; protecting the environment and providing environmental security; and increasing the forest / tree cover, there is a need to increase the availability of these from outside the natural forests.

4. Basic Objectives

The basic objectives of the National Agroforestry Policy are to:

- Encourage and expand tree plantation in complementarity and integrated manner with crops and livestock to improve productivity, employment, income and livelihoods of rural households, especially the small holder farmers.
- Protect and stabilize ecosystems, and promote resilient cropping and farming systems to minimize the risk during extreme climatic events.
- Meet the raw material requirements of wood based industries and reduce import of wood and wood products to save foreign exchange.
- Supplement the availability of agroforestry products (AFPs), such as the fuel-wood, fodder, non-timber forest produce and small timber of the rural and tribal populations, thereby reducing the pressure on existing forests.
- Complement achieving the target of increasing forest/tree cover to promote ecological stability, especially in the vulnerable regions.
- Develop capacity and strengthen research in agroforestry and create a massive people's movement for achieving these objectives and to minimize pressure on existing forests.

5. Strategy

Establishment of Institutional Setup at National Level to Promote Agroforestry

An institutional mechanism, such as a Mission or Board is to be established for implementing the agroforestry policy. It will provide the platform for the multi-stakeholders to jointly plan and identify the priorities and strategies, for inter-ministerial coordination, programmatic convergence, financial resources mobilization and leveraging, capacity building facilitation, and technical and management support. Such an institutional arrangement will ensure that agroforestry gets equal treatment with other agriculture enterprises, because at present whether in the sphere of inputs, markets, institutional finance, or research and extension, agroforestry is at a sub-optimal level. A suitable mechanism for coordination and convergence with state agriculture, and forest departments as nodal agencies may be established. The Mission / Board may be provided a corpus in order to effectively leverage Rs. 4000-5000 crores annually from the on-going programmes.

The Ministry of Agriculture has the mandate for agroforestry

Agroforestry Mission / Board will be located in the Department of Agriculture and Cooperation (DAC) in the Ministry of Agriculture (MoA). The Mission/ Board may comprise representatives of the Department of Agriculture Research & Education (DARE), Ministry of Environment & Forest (MoEF), Ministry of Rural Development (MoRD), Ministry of New and Renewable Energy (MN&RE), International Centre for Research in Agroforestry (ICRAF, South Asia Office), Planning Commission, National Rainfed Area Authority(NRAA), representatives of Non-Governmental Organizations, Industry, NABARD, Agricultural Universities, State Governments, etc. The actual implementation may involve convergence and dovetailing with a number of programmes. They may include, the Mahatma Gandhi National Rural Employment Guarantee Programme (MGNREGA), Integrated Watershed Management Programme (IWMP), National Rural Livelihood Mission (NRLM), National Bamboo Mission, Kisan Mahila Sashatikaran Pariyojana, Rashtriya Krishi Vikas Yojana (RKVY), National Medicinal Plants Board (NMPB), Mission for Integrated Development of Horticulture (MIDH), National Green India Mission, Warehouse Development and Regulation Act 2007 (WDRA), CAMPA fund. State Government may identify a Nodal Department for implementing the Agroforestry Mission/Board. At the district level, nodal agency may be Agriculture Technology Management Agency (ATMA) or any other department to be identified by the State, with other partners such as the Krishi Vigyan Kendras (KVKs), Van Chetna Kendras, Farmer's Associations, NGOs, private sector, Self-Help-Groups, Farmers' Cooperatives, Famer Producer Organizations, and Panchayati Raj Institutions (PRIs) etc. Agroforestry research and development (R&D), including capacity development and pilot studies / testing and action research should be the responsibility of the ICAR. ICRAF should be involved as an important partner in agroforestry research & development. Strength of Indian Council for Forestry Research & Education (ICFRE) should also be capitalized in this endeavor.

In the proposed institutional arrangement the current stakes of the key ministries are to be respected and utilized. It is envisaged that existing structures at the state, district and sub-district levels will be utilized with necessary strengthening. The important dimension of the implementation strategy is that it will effectively leverage tree-plantation, afforestation, greening, nutrition strengthening, etc., which are components of major flagship programmes to achieve its objectives. That will be its uniqueness. That will be its strength.

Simple regulatory mechanism

There is a need to create simple mechanisms / procedures to regulate the harvesting and transit of agroforestry produce within the State, as well as in various States forming an ecological region. There is also the need to simplify procedures, with permissions extended on automatic route as well as approval mode through a transparent system within a given time schedule. There are regulations imposed by multiple agencies of State governments (viz. Department of Forest, land revenue, other local bodies) on harvesting and transit which have negative implications on the growth of agroforestry. All these restricting regulations need to be identified and aligned with the proposed simplified mechanism.

Considering that States may find practical difficulties to exempt all tree species planted under agroforestry in farmland from State regulations for harvesting and transit, a practical way forward could be to develop a bouquet of commonly grown tress across the country and develop a uniform harvest and transit rules. The IVth Quinquennial Review Team (QRT, 2013) of NRCAF and AICRP on agroforestry has identified about twenty important multipurpose tree species at national level after thorough discussion with all the stakeholders, which can be the reference point while developing the bouquet.

A Committee constituted by the Ministry of Environment & Forests under Shri Arun Kumar Bansal has submitted its report in2011on the regulatory regime to be followed. The relevant recommendations of the Committee may be considered for inclusion in the implementation Guidelines for the National Agroforestry Policy.

Decentralized institutions of local governance, such as the PRI, GramSabhas, Joint Forest Management Committees (JFMCs), Eco Development Committees or other similar people's institutions, such as those under the Forest Rights Act (FRA), Panchayats (Extension to Scheduled Areas) Act 1996 (PESA) etc. may be considered for playing a role in the regulatory mechanisms. The National Agroforestry Policy should be consistent with the FRA, PESA and such Acts (viz. Chotanagpur Tenancy Act) under which such provisions of community rights for harvesting and transportation may have been already provided. A point of caution is that delegation of power to the decentralized institutions should be accompanied with the capacity building measures to equip them (including the Gram sabhas) with the knowledge on agroforestry rules and regulations, quality control, etc. to avoid possible misuse of the power.

Development of a sound database & information system

Security of land tenure is a critical issue. More than in other agricultural systems, due to the longer gestation period of trees relative to annual crops stability and security of tenure rights is a necessary condition for farmers' to take up agroforestry. In the absence of such rights farmers would be reluctant to invest their labour and meagre capital resources in a crop that may yield benefits after several years. A clear guarantee of tenure rights can support a farmer's strategy to invest in trees. Only then can farmers – as investors – make plans with confidence that the parameters shaping their long-term vision will not change. Similar is the case with the rights of the tenant farmers on trees planted in the farm which they have been cultivating or share-cropping for long. Tenancy rights must be respected. Even the financial institutions usually provide credit based on demonstration of security of tenure. Trees on farmland may be considered as collateral security for the purpose of financing. There is need for a more efficient system of data recording as at present there is complete lack of data on agroforestry. Number of trees on farm, species grown, survival rate, number of trees planted /harvested, or quantity of timber/ fuel wood/fodder produced is not available. The contribution of agroforestry to the national GDP is also not known. This poses a challenge for planning and implementation. However, in the land record system there is a provision of recording trees on farm. But, it is not clear whether the data recorded is authentic and used for any purpose. Updating the old land record systems with the help of modern geo-spatial tools (remote sensing and geographic information system), for recording of basic data such as existing trees, planting of new trees, and harvesting, etc. needs to be undertaken.

Investing in research, extension and capacity building and related services

There are over 30 research centers of the ICAR involved in agroforestry research, which is coordinated by the National Research Centre for Agroforestry (NRCAF). There are claims of availability of technology and appropriate agroforestry models suitable for different agro-climatic regions of the country. Also, advance research knowledge and facilities are reported to be available in the private domain, especially in the pulp and timber industry. However, there appears to be very little replication of such knowledge and models on the ground. The non-existent extension system for agroforestry may be the key reason for non-adoption of technologies; however, one needs to reconfirm the robustness of the technologies which have been developed. Therefore, it is suggested to set aside some seed money to fund time-bound research projects with specific objectives delivered through location specific basic, strategic and applied research.

Agroforestry cannot succeed without the willing support and cooperation of the people. It is essential, therefore, to inculcate in the people, a direct interest in agroforestry, its promotion and development. This can be achieved through the involvement of educational institutions, right from the primary stage. Farmers and interested people should be provided opportunities through institutions, such as the Krishi Vigyan Kendras, Trainers' Training Centres to learn agroforestry, agri-silvicultural and silvicultural techniques to ensure optimum use of their land and water resources. Short term extension courses and lectures should be organised in order to educate farmers. For this purpose, it is essential that suitable programmes are propagated through mass media, audio -visual aids and the extension machinery. There is need to develop yield and volume tables of agroforestry tree species. Likewise, the species with high carbon sequestration capacity need to be identified. Also, a common web-based platform may be setup to bring all research findings and available knowledge in the area of agroforestry for access of all stakeholders.

Improving famers' access to quality planting material

Certification of nurseries, seeds and other planting materials for agroforestry is required to make available good quality planting material at the required scale. Institutional mechanism for registration of nurseries and their accreditation should be established. Private sector can play an important role in augmenting supplies of improved planting stock as demonstrated in case of poplar and clonal eucalyptus plantations. Hence, the private sector participation should be encouraged in production and development of supply chain of quality planting materials.

Providing institutional credit and insurance cover for agroforestry

There is need for setting up of special purpose vehicles /banking institutions to address specific needs of agroforestry sector. Agroforestry sector should also be benefitted with the provisions of interest subvention in the line of agricultural credit. Dedicated Farmers Producers Organizations (FPO) be promoted to organize the farmers and take up agroforestry at economies of scale.

Facilitating increased participation of industries dealing with agroforestry produce

The role of agroforestry/biomass based industries in the promotion of agroforestry is crucial. Therefore, such industry sector needs to be encouraged and facilitated. The role of industries in the promotion of agroforestry can be tapped in multiple ways, especially in the areas of (a) production and supply chain development for high quality planting materials, (b) technology development and dissemination, especially for planting materials, processing, etc. (c) providing extension services to the farmers, (d) providing market information and future trends, (e) certification of nurseries, seeds and finished products for sustainable management practices, (f) developing agroforestry plantation on government lands on lease contract and in partnership with local people's institutions, etc. Policies are therefore, required to recognize the role of industries and make enabling conditions for increasing the participation of the industries. Agroforestry be treated as a priority area under Corporate Social Responsibility programmes.

Strengthening farmer access to markets for tree products

Barring few sporadic instances, the marketing system for agroforestry produce is unorganized. Marketing infrastructure similar to what is available for agricultural commodities, including market information be introduced with more private sector participation.

Incentives to farmers for adopting agroforestry

Regional and thematic differentiation in the agroforestry policy needs to be minimized. Incentive and support structure, such as the input subsidy, interest moratorium, etc. during the gestation period for promotion of agroforestry be provisioned to encourage farmers. Clear directives for value chains development need to be incorporated for the agroforestry sector over a period of time. The role of Farmer Producer Organizations (FPOs) in the promotion and value chain development for agroforestry should be considered.

Promoting sustainable agroforestry for renewable biomass based energy

Emphasis needs to be on raising fast growing trees / bushes / grasses on marginal and degraded farmlands keeping in view their uses for meeting various energy requirements for making profitable agroforestry practices. This is of particular significance for meeting various energy needs of agroforestry itself, such as for irrigation, motive power, farm machines and processing industry. Therefore, it should be considered in conjunction with making of the provisions for financial incentives, especially for setting up of various renewable energy systems / devices.

6. Pathway for achieving Policy Deliverables

- Mainstreaming agroforestry in agriculture policies and strategies.
- A dedicated corpus be created to leverage resources available under various schemes/programmes/missions in undertaking focused and synchronized interventions for agroforestry sector particularly in meeting the gaps and up-scaling the efforts in a coordinated manner.
- States to create enabling environment and legislation and simplify regulations related to forestry, land use & land tenure, especially those linked to harvesting and transportation of trees grown on farms.

- States have to identify about 20 commonly grown trees species which can be grown on farmlands for the economic and ecological benefits of the farming community. These species have to be notified for exemption from any state regulatory regime, especially on growing, harvesting and transit.
- States to ensure a secured land tenure system, safeguarding the interest of small and marginal farmers and create a sound base of land records and data for developing an MIS for agroforestry for a transparent and non-controversial operational system.
- Public private partnership (PPP) to be encouraged for road side/canal side/barren community land/other non-forest waste lands for promotion of agroforestry to provide opportunities of economic returns and contributing ecological services.
- Providing quality and certified planting material, at local level through promotion of nurseries, duly registered and accredited by a third party, by involving government/private sector.
- Data collection with source of agroforestry produce at National level by recognized statistical organizations (viz. CSO, NSSO) to be done to have legality data of source of agroforestry produce to facilitate hassle free harvesting/transport/traceability of source/chain of custody.
- Agroforestry research to be encouraged, both in government and private sector, particularly for multipurpose indigenous species with higher nitrogen-fixing ability, so as to meet the local needs for fuel, fodder and timber as well as improving the soil health. It should also focus on developing market driven models suitable to different ecological conditions to encourage farmers for adopting agroforestry as a viable enterprise.
- National Research Centre for Agroforestry (NRCAF) may be upgraded to a National level Institute of Agroforestry with regional setups in major agro-climatic zones of the country. Agroforestry research wing of ICFRE also be strengthened and taken advantage of to provide stimulus and create an enabling environment for the growth of private research and extension services.
- Appropriate extension mechanism equipped with scientific setup involving State Agriculture Universities (SAUs), Krishi Vigyan Kendras(KVKs), Van Chetna Kendras etc. to be put in place for agroforestry. Cost-effective extension models may be devised involving farmer's groups, NGOs, public/private agencies, Farmer Producer

Companies, etc. to disseminate knowledge/information of this sector. Integrating agroforestry content in the agriculture extension packages and developing an unified extension system for all farming systems in the country.

- Encouraging agroforestry as a course curriculum in school education and motivating youths to grow and conserve trees.
- National Bureau for Plant Genetic Resources (NBPGR) to focus on conserving, monitoring and providing guidelines for germplasm exchange of agroforestry species.
- Marketing infrastructure including market information system to be put in place with active collaboration of private sector. Contract farming, Public Private Partnership, Special Purpose Vehicles mechanisms may also be explored to promote and upscale agroforestry. Road side/ canal side/barren community land/other non-forest waste lands to be encouraged for plantation of agroforestry tree species to provide opportunities of economic returns as well as contributing towards ecological benefits. These activities may be promoted through public private partnership mode.
- Agroforestry farmers also to be considered eligible for incentives on input subsidy, post-harvest management facilities, interest moratorium etc. as are being provided to farmers growing agricultural crops.
- Specific products/ special purpose vehicles may be devised to meet the credit and insurance needs of agroforestry sector. Interest subvention in the line of agricultural credit be extended to agroforestry sector. Agroforestry commodities also be enlisted under Warehouse Development and Regulation Act 2007 (WDRA) for ensuring adoption of quality standards of the "Warehousing Manual for Operationalizing of Warehousing (Development and Regulation) Act, 2007 so as to become eligible for availing finance for harvested produce of agroforestry.
- To create an enabling environment to implement strategies for quantifying carbon sequestration and other environmental services for the economic benefit of farmers.
- Industries to be encouraged as end user for promotion of agroforestry

produce, value chain development, technology development and market information etc.

References

CGIAR.2015.CGIAR Strategy and Results Framework 2016-2030.http://www.cgiar.org/our-strategy/.Critical political ecology: The politics of environmental science. Routledge. PawsonR.2002.Evidence-basedpolicy: Thepromiseofrealistsynthesis. Evaluation, 8(3), pp.340-358.

Forsyth T. 2004. Adhar-Schuster M, Thomas RJ, Stringer LC, Chasek P, Seely M. 2011. Improving the enabling environment to combat land degradation: Institutional, financial, legal and science- policy challenges and solutions. Land Degradation & Development,22(2), pp.299-312.

Plant M. 2003. Evidence-based policy or policy-based evidence? Addiction, 98, pp.397-411.

Sabatier PA. 1987. Knowledge, policy-oriented learning, and policy change an advocacy coalition framework. Science Communication, 8(4), pp.649-692.

Sanderson I. 2002. Evaluation, policy learning and evidence-based policy making.Public Administration, 80(1), 1-22.

TomichTP, ChomitzK, FranciscoH, mIzacAMN, MurdiyarsoD, RatnerBD, ThomasDE, van Noordwijk M. 2004. Policy analysis and environmental problems at different scales: asking the right questions. Agriculture, Ecosystems & Environment, 104(1), 5-18.

Tripp R. 2003.The enabling environment for agricultural technology in sub-Saharan Africa and the potential role of donors. Overseas Development Institute, London (UK).van Noordwijk M, Coe R, Sinclair F. 2016. Central hypotheses for the third agroforestry paradigm within a common definition. World Agroforestry Centre (ICRAF) Working paper 233, Bogor (Indonesia). http://dx.doi.org/10.5716/WP16079.PDF

10

Horticultural Policies and Development in India

Sanghamitra Pattnaik

KVK, Mayurbhanj 1, Odisha University of Agriculture & Tehnology, Bhubaneswar

1. Introduction

India's horticulture sector stands as a beacon of promise and potential within the agricultural landscape, showcasing a rich diversity of fruits, vegetables, tubers, spices, flowers, medicinal and ornamental plants as well as herbs. This sector provides livelihoods for millions of farmers and plays a crucial role in ensuring food security and nutritional well-being. Amidst the challenges posed by climate change and environmental degradation, the resilience of horticultural crops becomes increasingly significant. Diversifying towards resilient crops and sustainable practices not only mitigates risks but also enhances the sector's adaptability to changing climatic conditions. Investment in research and development to breed climate-resilient varieties and promote agroecological approaches is imperative. Furthermore, promoting organic and niche horticulture products presents exciting opportunities for Indian farmers to capture premium markets both domestically and internationally. With growing consumer awareness and demand for healthy and nutritionally rich food, India's horticulture sector is well-positioned to become a global leader in organic and exotic produce.The recent trend of exploring new crops within Indian horticulture promises to diversify farmers' income sources and address emerging challenges such as climate change and shifting consumer preferences. The cultivation of exotic fruits and vegetables, which fetch premium prices and offer resilience against climate variability, represents a paradigm shift in Indian horticulture, opening up a world of possibilities for farmers and consumers alike. By embracing diversity and innovation, we can build a resilient and sustainable agricultural system that nourishes both people and the planet. The technological aspects of cultivating new varieties, growing crops in non-traditional areas, plant protection, floral phenology, and landscaping native ornamentals will facilitate the adoption of new and non-traditional crops, inspiring Indian farmers to explore and develop new tastes and preferences.

The landscape of Indian agriculture has undergone a significant transformation in recent years, with horticulture emerging as a key driver of prosperity and sustainability. Over the past two decades, the horticulture sector has experienced remarkable growth, nearly doubling its production to 320.77 million tonnes in 2019-20. This surge in output has outpaced the growth rate of food grain production, highlighting the crucial role of horticulture in driving agricultural diversification and income augmentation. Recognizing its potential, the Government of India has made substantial investments, allocating Rs 2,250 crore during the fiscal year 2021-22 to propel the growth trajectory of the horticulture sector.

Policies and phases of horticultural development: The development of horticulture in India can be broadly categorized into several phases, each marked by significant milestones and shifts in policies, technology, and market dynamics. During the pre-independence period, horticulture was largely practiced using traditional methods, with a focus on subsistence farming and limited commercial cultivation. Cultivation was restricted to local varieties of fruits, vegetables, and flowers, resulting in minimal diversity.

From the post-independence era to the Green Revolution (1947-1960s), the focus remained on achieving food security, and horticulture was not a priority. However, the establishment of agricultural universities and research institutions during this period laid the groundwork for future horticultural development.

During the Green Revolution and beyond (1960s-1980s), the primary focus was on increasing cereal production, although horticulture began to gain some attention with the introduction of high-yield varieties and improved practices. The Indian Council of Agricultural Research (ICAR) and state agricultural universities initiated specific programs for horticultural crops, leading to gradual advancements in the sector.

The period from the 1980s to the 2000s saw significant progress with government initiatives like the National Horticulture Mission (NHM) in 2005, aimed at the holistic growth of the horticulture sector. This era introduced advanced technologies such as tissue culture, protected cultivation, micro-irrigation, and integrated pest management. There was also a focus on post-harvest management, cold chain infrastructure, and marketing facilities. During the 1980s, the rural economy experienced significant diversification, marked by a rapid increase in the production of non-food grain commodities such as milk, fish, poultry, vegetables, and fruits. This diversification contributed to a notable acceleration in agricultural GDP growth during this period (Chand,

2003). Liberalization policies from the 1990s to 2010 opened up new markets and opportunities for export, with increased involvement of the private sector in horticulture, leading to better input supply, processing, and marketing.

The most recent phase (2000s-present) saw the launch of the Mission for Integrated Development of Horticulture (MIDH) in 2014, consolidating all previous schemes under one umbrella for comprehensive development. This period emphasized improving the quality of produce to meet global standards and boosting exports. The adoption of digital tools and precision farming techniques enhanced crop management and yield prediction by leveraging data-driven insights and advanced technologies to optimize horticultural practices. There was also a strong emphasis on organic farming, sustainable practices, and climate-resilient horticulture.

Current trends and future prospects in horticulture include a growing interest in rooftop gardening, vertical farming, and peri-urban horticulture to meet urban demand. There is a focus on value-added products such as processed fruits, vegetables, tubers and floriculture products. Ongoing research aims to develop new varieties, pest-resistant strains, and sustainable farming practices. Continued government support through subsidies, schemes, and programs aims to enhance productivity and profitability in horticulture.

The phases of horticultural development in India reflect a journey from traditional subsistence farming to technology driven advanced, market-oriented sector. Each phase has contributed to making horticulture a significant part of Indian agriculture, ensuring food security, enhancing farmer incomes, and contributing to the economy through exports.

Challenges in the Horticulture Sector: Despite its transformative potential, the horticulture sector faces challenges, particularly in areas such as post-harvest infrastructure and value addition. The absence of Minimum Support Price (MSP) for perishable horticultural products, coupled with inadequate transportation and storage facilities, hampers the sector's profitability and market access. Strategic interventions focusing on research and development, resource optimization, and value addition can unlock the latent potential of horticulture. Value addition, through processing and packaging, enhances the economic value of horticultural produce, fosters nutritional security, and opens new market avenues. Integration of modern machinery and technology-driven solutions is key to enhancing efficiency and productivity in horticultural management.

Horticultural Research in India and its impact: This began systematically in 1954 with the establishment of the Division of Botany at Indian Council of Agriculture Research (ICAR)-Indian Agricultural Research Institute (IARI), focusing on fruit, vegetable, and ornamental crops. Over the years, numerous independent institutions, including 10 Central Institutes, 6 Directorates, 7 National Research Centres, 13 All India Coordinated Research Projects, and 6 Network Projects, have been launched to support the national horticulture agenda. Additionally, several crop and commodity boards, such as the National Horticulture Board and various Research and Development(R&D) establishments like Council of Scientific & Industrial Research (CSIR), Department of Biotechnology (DBT), Defence Research and Development Organisation (DRDO), Indian Institute of Technologies (IITs), and Bhabha Atomic Research Centre (BARC), have contributed significantly to the horticultural field. These collaborative efforts have led to the development of improved crop management practices, new and resilient crop varieties, comprehensive plant protection schedules, and advanced post-harvest technologies. Together, these innovations have enhanced productivity, quality, and sustainability in Indian horticulture, positioning the country as a prominent player in the global horticultural sector.

Research in horticulture has significantly advanced the sector through several key achievements. Advanced propagation techniques, such as tissue culture in banana, shoot tip grafting, micro-grafting, and micro-budding in citrus and potato aeroponics, have enabled large-scale production of disease-free planting materials. Canopy management practices, high-density planting, and rejuvenation technology for old orchards have been standardized, increasing productivity in various fruit crops. The development of hybrid and disease-resistant vegetable varieties by ICAR, has improved environmental safety and crop resilience. Innovations in micro-irrigation, fertigation, and nutrient management have enhanced water and nutrient use efficiency, crucial for drought mitigation. Additionally, genetic improvements have addressed abiotic stresses and enabled cultivation in arid and saline conditions. Good Agricultural Practices (GAP) and advancements in farm mechanization have reduced labour intensity and improved efficiency. Post-harvest handling protocols and value-added products have supported export promotion and on-farm storage.

Amrapali, a unique mango hybrid (Dashehari x Neelum) released in 1979, is known for its precocity, dwarf stature, high yield, and suitability for high-density planting due to its positive response to pruning. Covering over two hundred thousand hectares across India, it is notably prevalent in Odisha with seventy-three thousand hectares in districts such as Keonjhar, Dhenkanal, Angul, and Kashipur Block of Raygada (A.K. Singh et al, 2022). Similarly, the Dogridge grape rootstock (Vitis champinii), developed to withstand drought and salinity, enhances yield and quality in commercial varieties. Mahatma

Phule Krishi Vidyapeeth, Rahuri's Bhagawa and Super-Bhagwa pomegranate varieties outperform existing types like Ganesh and Arakta in yield and quality, with Bhagawa yielding heavier fruits (320.6 g) with a brighter red rind compared to Ganesh (297.8 g). In bananas, biotic and abiotic stress-tolerant varieties like Kaveri Haritha (Fusarium wilt) and Kaveri Saba (drought and saline soils) have boosted production. The Arka Rakshak hybrid tomato, resistant to multiple diseases, offers significant yield potential (19 kg/plant; 100 t/ha). Potato varieties such as Kufri Jyoti, Kufri Bahar, Kufri Pukhraj, and Kufri Chipsona 1 dominate 75% of India's potato cultivation.

Recent developments include tomato hybrids (Arka Apeksha and Arka Visesh) for ketchup and paste, the dehydration-suitable onion variety Arka Yogith, French fry-specific potato variety Kufri Frysona, curcumin-rich turmeric variety IISR Pragati, pomegranate variety Solapur Anardana for anardana, and grape variety Manjari Medika for juice. The Arka Prajwal tuberose, with its bold white flowers tinged with pink, and turmeric varieties IISR Sudarshana (7.9% curcumin), IISR Prathibha (6.2%), and IISR Alleppey Supreme (6.0%) offer enhanced consumer value over older varieties like Suvarana (4.0%) and Suguna (4.7%). Cassava (tapioca) promising 'Sree Kaveri' variety , resistant to cassava mosaic disease, starch content 27-28%, 9-10 months old and average yield of 40-50 t/ha ; 'Sree Sakthi' variety ,completely resistant to cassava mosaic disease, industrial variety & starch content 27-28%, 9-10 months old and average yield of 40-45 t/ha and 'Sree Suvarna' variety , completely resistant to cassava mosaic disease, starch content 25-27%, 7-8 months old and average yield of 35-40 t/ha have been developed by ICAR-CTCRI scientists. Similarly they have developed Sweet Potato 'Bhu Sona' variety (yellow skin & dark orange flesh, beta carotene content 13.2-14.4 mg/100 g fresh weight, duration 105-110days, yield potentiality20-24 t/ha); 'Bhu Swami' variety (105-110 days duration , extractable starch 21 % & yield potentiality of 20t/ha) and 'Bhu Krishna' sweet potato variety (dark purple skin & dark purple flesh, highly tolerant to sweet potato weevil, Anthocyanin content 90 mg /100 mg fresh weight , 110-120 days duration & yield potentiality of 18-22 t/ha). Promising Taro variety 'Sree Hira', tolerant to leaf blight disease, bears 12-16 cormels per plant, produces yield of 16-20 t/ha, harvesting at 180 days after planting and another one 'Sree Telia' harvesting at 120 days after planting& produces yield of 10-12 t/ha have been developed during 2023 by ICAR-Regional Centre, Central Tuber Research Institute, Odisha.

Indian Minimum Seed Certification Standards for Horticultural Crops: In the realm of agriculture, the assurance of quality seeds stands as the bedrock for a bountiful harvest, echoing the sentiment that a good seed heralds a promising yield. Rooted in the foundation laid by the Seeds Act of

1966, India's systematic endeavour towards seed certification has evolved into a pivotal component of the nation's agricultural landscape. Through the concerted efforts of State Seed Certification Agencies, and bolstered by the guiding hand of the Indian Council of Agricultural Research (ICAR) and the National Agricultural Research and Education System (NARES), the Indian Minimum Seed Certification Standards have emerged as a beacon of quality assurance, ensuring genetic purity and identity of varied crop varieties. As the agricultural canvas expands with the introduction of new crop varieties, particularly in the realm of horticulture, the recent publication of the second volume of the 'Indian Minimum Seed Certification Standard' stands as a testament to India's commitment towards enhancing the quality of its agricultural produce. By bridging the gap for crops previously omitted from certification standards, this publication not only integrates diverse varieties into the certified seed production chain but also propels the nation closer to the vision of 'Aatma Nirbhar Bharat-Krishi' (Self-Reliant India-Agriculture). With its comprehensive coverage and meticulous standards, this volume is poised to empower seed production and certification agencies, catalysing a paradigm shift towards sustainable agriculture and bolstering the resilience of Indian farmers against the vagaries of nature.

Advancements in Fruit Production Techniques: The evolution of fruit production techniques has been instrumental in enhancing agricultural productivity and meeting the growing demand for nutritious fruits. In recent years, there has been a significant focus on improving fruit cultivation technologies to maximize yield, quality, and sustainability. This shift is driven by the increasing recognition of the importance of fruits in a balanced diet and the rising consumer demand for fresh, high-quality produce. As a result, extensive research and development efforts have been directed towards advancing fruit production methods to address various challenges faced by growers.One of the key areas of innovation in fruit production is the adoption of modern technology and machinery. Mechanization has revolutionized agricultural practices, allowing for increased efficiency and productivity. From the use of combine harvesters and tractors to precision farming tools and drones, modern technology is reshaping the way fruits are cultivated, harvested, and managed. These advancements enable farmers to optimize resource utilization, minimize labour requirements, and enhance overall crop yields.Another significant advancement in fruit production is the integration of artificial intelligence (AI) and data-driven technologies. AI-powered tools, such as weather prediction systems and remote sensors, provide farmers with real-time insights into environmental conditions, pest outbreaks, and crop health. By leveraging AI-driven analytics, growers can make informed decisions regarding irrigation, fertilization, and pest management, thereby improving crop resilience and productivity. Biotechnology has also played a crucial role in advancing fruit production techniques. Through genetic engineering and biotechnological interventions, researchers have developed resilient crop varieties with enhanced resistance to pests, diseases, and environmental stressors. These genetically modified fruits offer growers a sustainable solution to combatting agricultural challenges while ensuring consistent yields and quality.Furthermore, innovative practices such as high-density planting, meadow orcharding, and the use of dwarfing rootstocks have emerged as effective strategies for maximizing fruit production in limited spaces. These techniques optimize land utilization, increase planting density, and promote early fruiting, resulting in higher yields per unit area. Additionally, advancements in water harvesting, soil management, and disease control have further contributed to the sustainability and profitability of fruit cultivation.

Strategies for Sustainable Growth in India's Vegetable Agriculture Sector: In strategizing fresh vegetable production, it's crucial to acknowledge the inherent opportunity the vegetable agriculture sector provides in tackling the interconnected issues of food security and nutritional welfare. India, with its diverse climate conducive to cultivating a wide array of vegetable varieties, has emerged as a global leader in vegetable production. However, the sector faces formidable obstacles such as suboptimal productivity, inadequate infrastructure, and the evolving impacts of climate change. To navigate these challenges and unlock the sector's full potential, concerted efforts are essential across the realms of research, policy formulation, and on-the-ground implementation. Innovative approaches to boosting productivity are at the core of sustainable vegetable agriculture. Institutions like the Indian Council of Agricultural Research (ICAR) and Agricultural Universities (AUs) have been pivotal in developing high-yielding varieties and hybrids tailored to India's diverse agro-climatic conditions. Embracing cutting-edge technologies such as precision farming, drip irrigation, and greenhouse cultivation holds tremendous promise in optimizing resource utilization and enhancing productivity. Furthermore, strategic investments in research aimed at developing climate-resilient varieties are crucial for bolstering the sector's capacity to withstand the increasingly erratic weather patterns brought about by climate change. Ensuring quality for domestic consumption and global trade is paramount to meet the burgeoning demands of both domestic consumers and international markets. Initiatives aimed at promoting adherence to Good Agricultural Practices (GAP) standards, bolstering cold chain infrastructure, and widespread adoption of biofertilizers and biopesticides play a pivotal role in enhancing the overall quality assurance framework. By upholding stringent quality standards, Indian vegetable producers can not only cater to the growing domestic market but also tap into lucrative export avenues, thereby boosting the nation's economy.At the heart of the vegetable agriculture sector are the farmers, whose resilience, ingenuity, and hard work drive progress. Empowering farmers through targeted capacity-building initiatives, including comprehensive training programs, knowledge-sharing platforms, and enhanced access to credit facilities, is critical to enabling the adoption of modern agricultural practices. Moreover, fostering inclusive growth by integrating smallholder farmers into value chains and promoting gender equality in agriculture can unlock the sector's full economic potential while ensuring equitable distribution of benefits.The pivotal role of vegetable crops cannot be overstated, as India charts its course towards a more sustainable agricultural future. By harnessing the transformative power of science, technology, and innovation, India can not only meet the nutritional needs of its burgeoning population but also emerge as a global leader in vegetable production and trade.

Progress and Prospects of Medicinal and Aromatic Plants in India: Medicinal and aromatic plants (MAPs) have stood as pillars in human healthcare for centuries, offering a rich tapestry of remedies ranging from herbal teas to pharmaceutical compounds. Despite advancements in modern medicine, approximately 80% of the global population still relies on traditional plant-based remedies for primary healthcare needs, as highlighted by the World Health Organization. The cultivation of these plants has evolved significantly, transitioning from small backyard gardens to large commercial plantations, driven by advancements in agricultural technology, increasing demand for plant-based medicines, and the necessity to preserve endangered species. Cultivating medicinal plants involves a nuanced interplay of factors including species selection, optimizing growth conditions, and managing pests and diseases. The significance of medicinal plants extends beyond healthcare, infiltrating the realms of global trade. The estimated value of the global herbal industry exceeds US$ 60 billion, encompassing pharmaceuticals, spices, herbs, natural cosmetics, and essential oils. With an annual growth rate of 7%, this industry is poised to reach US$ 5 trillion by 2050 (Manish Das,2023). India has emerged as a key player in this global trade, although realizing its potential necessitates a comprehensive strategic plan to bolster exports and minimize imports. However, the increasing demand for MAPs has raised concerns regarding the conservation of genetic diversity, habitat loss, and product quality. Ensuring the future sustainability of the medicinal and aromatic plants sector requires a multifaceted approach. This entails continuous efforts to secure quality raw materials, adhering to standard procedures such as good agricultural and collection practices, laboratory practices, and manufacturing practices. It also necessitates the development of new varieties, adoption of good agricultural practices, and the implementation of quality assessment methodologies utilizing cutting-edge technologies. Organizations like the ICAR-Directorate of Medicinal and Aromatic Plants Research (ICAR-DMAPR) play a pivotal role in this endeavour through coordinated research and development programs. With achievements including the development of numerous varieties and technologies benefiting farmers and stakeholders, ICAR-DMAPR stands at the forefront of advancing the medicinal and aromatic plants sector. This special issue serves as a comprehensive compilation, consolidating scattered information to shed light on this vital aspect of global healthcare and trade.

Advancements in Indian Floricultural Exports: The journey of India's floriculture sector from traditional pushcart transportation to modern chartered flights is truly remarkable, positioning India as a key player in the global market with major importing countries including the USA, Netherlands, UAE, UK, Canada, and Malaysia. Encompassing a diverse range of products such as cut flowers, pot plants, cut foliage, seeds, bulbs, tubers, rooted cuttings, and dried flowers or leaves, the sector is dominated by key flowers like roses, carnations, chrysanthemums, and orchids. A notable innovation is the advanced packing technology for jasmine flowers, utilizing pharmocol packaging with gel ice and aluminium foil lining, which has reduced post-harvest losses from 30% to 10%, extended shelf life up to 72 hours, and preserved colour. This technological advancement has significantly enhanced the export potential of jasmine flowers. The liberalization of industrial and trade policies has encouraged the establishment of export-oriented floriculture units, leading to significant growth. There are more than 300 export-oriented units in India. More than 50% of the floriculture units are based in Karnataka, Andhra Pradesh and Tamil Nadu. With the technical collaborations from foreign companies, the Indian floriculture industry is poised to increase its share in world trade. The evolution of India's floriculture sector showcases the synergy between traditional knowledge and modern technology, contributing to the economy and enriching global floral markets with further potential for growth and innovation.

India's Mushroom Production: Presently, India contributes approximately 0.4% to the total global production of mushrooms. Over the past decade, the country's mushroom productivity has nearly doubled, with production experiencing a remarkable increase by more than 36 times. The Indian Council of Agricultural Research (ICAR) has played a pivotal role in this growth by releasing several promising varieties of cultivated mushrooms. These include varieties of button mushroom, paddy straw mushroom, shiitake mushroom, milky mushroom, oyster mushroom, and macrocybe mushroom. Among these, the first non-browning button mushroom variety, DMR-NBS-5, has gained widespread acceptance across the country, now accounting for 32% of India's total button mushroom production.

Export Performance of Indian Fruits, Vegetables, and Floriculture in 2023-24: During the fiscal year 2023-24, the export of fresh fruits and vegetables from India showed significant volumes and values as reported by APEDA. Fresh onions led the category with an export volume of 17.17 lakh metric tons (MT) valued at Rs. 392,278.32 lakhs. Fresh vegetables followed with 9.18 lakh MT (Rs. 293,827.21 lakhs), while fresh mangoes and grapes were also notable contributors, with exports of 0.32 lakh MT (Rs. 49,546.43 lakhs) and 3.43 lakh MT (Rs. 346,070.22 lakhs) respectively. In the processed fruits and vegetables segment, cucumber and gherkins (prepared and preserved) reached 2.44 lakh MT (Rs. 212,707.32 lakhs), and processed vegetables totalled 5.37 lakh MT (Rs. 652,346.97 lakhs). Mango pulp exports amounted to 0.60 lakh MT, valued at Rs. 62,429.33 lakhs. The floriculture and seeds sector also demonstrated noteworthy exports, with floriculture products at 0.19 lakh MT (Rs. 71,782.88 lakhs) and fruits and vegetable seeds at 0.14 lakh MT (Rs. 100,495.92 lakhs). These figures highlight the robust performance and global demand for India's agricultural exports in various categories.

Role of Horticulture in Agroforestry: Horticulture is a fundamental component of agroforestry systems, significantly enhancing their multifunctionality and sustainability. By incorporating the cultivation of fruits, vegetables, nuts, and ornamental plants with traditional forestry and agricultural practices, horticulture offers diverse economic, environmental, and social benefits. Socially, horticulture enhances food and nutritional security, supports cultural practices, and provides educational and recreational opportunities. Examples include home gardens, silvi-horticulture systems, agro-silvo-pastoral systems, and urban/peri-urban agroforestry. Despite challenges like the need for training, market access, and research, horticulture remains vital for the multifunctionality and sustainability of agroforestry systems.

Government Initiatives and Investments: Mission for Integrated Development of Horticulture (MIDH) Mission for Integrated Development of Horticulture (MIDH) is a Centrally Sponsored Scheme for the holistic growth of the horticulture sector covering fruits, vegetables, root & tuber crops, mushrooms, spices, flowers, aromatic plants, coconut, cashew, cocoa and bamboo. Under MIDH, Government of India (GOI) contributes 60%, of total outlay for developmental programmes in all the states except states in North East and Himalayas, 40% share is contributed by State Governments.

In the case of North Eastern States and Himalayan States, GOI contributes 90%. In case of National Horticulture Board (NHB), Coconut Development Board (CDB), Central Institute for Horticulture (CIH), Nagaland and the National Level Agencies (NLA), GOI contributes 100%. MIDH also provides technical advice and administrative support to State Governments/ State Horticulture Missions (SHMs) for the Saffron Mission and other horticulture related activities. National Horticulture Mission (NHM) launched during 2005-2006 is one of the sub scheme of Mission for Integrated Development of Horticulture (MIDH) .NHM aims at the holistic development of the horticulture sector covering fruits, vegetables, root & tuber crops, mushrooms, spices, flowers, aromatic plants, coconut, cashew and cocoa by ensuring forward and backward linkage through a cluster approach with the active participation of all stakeholders. Major interventions under NHM include, supply of quality planting material through establishment of nurseries and tissue culture units, production and productivity improvement programmes through area expansion and rejuvenation, technology promotion, technology dissemination, human resource development, creation of infrastructure for post-harvest management and marketing in consonance with the comparative advantages of each state/region and their diverse agro-climatic conditions. Horticulture Mission for North East and Himalayan states (HMNEH) aims to achieve overall development of Horticulture in NE and Himalayan states. The mission covers all NE States including Sikkim and three Himalayan states of Jammu & Kashmir, Himachal Pradesh and Uttarakhand.

Unlocking the full potential of horticulture necessitates the collaborative efforts of all stakeholders along the value chain. Research-led innovation, coupled with policy support and infrastructural investments, can propel the sector towards sustainable growth and resilience. Promoting pollinator-friendly practices and diversification strategies can further enhance the sector's productivity and resilience to climate variability. With the government's renewed focus and initiatives aimed at accelerating horticultural development, the sector is poised for a paradigm shift towards prosperity and sustainability.

Conclusion

The landscape of Indian horticulture has undergone a significant transformation, transitioning from traditional subsistence farming to technology driven advanced, market-oriented sector. With continuous advancements in technology, government support, and strategic investments in research and development, the horticulture sector is poised for sustainable growth and resilience. By embracing diversity and innovation, India can build a resilient and sustainable agricultural system that nourishes both people and the planet.

References

AK Singh, Murthy BNS and E Sreenivasa Rao. ICAR(2022), Indian Agriculture after independence-Achievements in Horticulture in Independent India, 6th chapter, pp114-133

Apeda.gov.in

Bhat MH (2017) Development of Horticultural Sector in India – The Way Forward. J Agroecol Natural Resource Manage 4(4): 292-297.

Manish Das, Director & Project Co-ordinator (2023), Medicinal and Aromatic Plants- Key for health, Indian Horticulture, Sept-Octo,2023, Vol68, N05, pp3

Rabindra Verma (2022) , Horticulture Sector- Opening up new avenues , Indian Horticulture, Jan-Feb,2022,Vol67,No1

Ruprekha Buragohain, Assam Agricultural University, Jorhat(2024), Recent advances in production Technology of Fruit crops , Article ID 45, Just Agriculture, Multidisciplinary e-Newsletter (e-ISSN: 2582-8223), Vol4,Issue 5,January 2024

S.K. Malhotra (2023), Indian Minimum Seed Certification Standards for Horticultural Crops, Indian Horticulture ,July- Aug , 2023, Vol 68, No4,pp2

S.K.Malhotra(2023) , Strategizing fresh vegetable fresh vegetable production, Indian Horticulture , Mar-April, 2023, Vol 68, No2, pp2

V.P.Patel & P.C.Tripathi (2023), Scenario of Horticulture Research and development in India and for northern India, Indian Horticulture, Nov-Dece,2023, Vol68,No6, pp6-10

11

Land Tenure and Agricultural Productivity

Vivek Pattanayk

Retd.s IAS

The land is one of the basic natural resources for agricultural growth and development of a country. Investment on agricultural land improvement by way of land leveling, contour bunding, sinking of well for irrigation and resorting to other measures like treating the soil to reduce acidity etc. have been continuous effort of agriculturists for increasing productivity.In addition, new variety of seeds which are high yielding and pest resistant are sought after by farmers for increasing production. Efficient land management encompassing soil and water conservation is the building block for sustainable agricultural development.

Land tenure rules define the ways in which property rights to land are allocated, transferred, used, or managed in a particular society. Ownership of land plays a significant role for improvement of land and its management. If the land is tenanted,the tiller of the land has less stake on long term measures on land.For centuries, the land tenure system in India was based on intermediaries known as Zamindary system.In Odisha, the tenurial system has based on the Bengal Rents Act, the Bengal Tenancy Act, the Bihar & Orissa Tenancy Act and finally, the Orissa Tenancy Act.In addition, theCentral Province Tenancy Act and the Central Province Land Revenue Act were in vogue in the Sambalpur district when the Orissa province was formed. Further, theGanjam district was under the Madras Estates Land Act(MELA). All these laws defined and classified the tenurial system.If one examines the OT Act one would find series of intermediaries between the proprietor and actual tiller of land.Tenure-holders had under-tenure holders. Raiyats had under-Raiyats. Tenants also inducted sub-tenants.

Only after India became independent intermediaries were abolished, and that too over along period of time. There were two main objectives relating to abolitionof intermediaries. One was economic and second was social. Unless the tiller of land is given right over the land, he would not take interest in its development and productivity would not increase. The social objective was to dilute concentration of wealth. While economic reason had easy acceptability among the policy makers and planners, social objective was confronted at multiple levels.

The Orissa Estate Abolition Act came into force in1953.This statute was epoch-making law as it abolished zamindary system in the State.Abolition of estates was not an easy matter as the laws relatingto their abolition and compensation were contested in the judicial fora and these cases went right up tothe Supreme Court who gave adverse decision declaring them as unconstitutional violating fundamental rights in particular the Right to Property as outlined under erstwhile Article 31 under the Constitution. In order to protect these legislations amendment to the Constitution had to be done. A special schedule to the constitution was incorporated.

With the abolition of intermediaries,ownership did not automatically vest on the tillers of land.Big land holders who were owners continuedto induct tenants to cultivate land.Many of them were absentee landlords.In Odisha in gradual manner tenancy reforms were made by way of legislation of the Orissa Tenancy Reforms Act and the Orissa Tenants Protection Act.Finally Land Reforms Act was introduced in 1960.

Under this land reforms law first land ceiling of ownership was imposed and second it discouraged induction of tenants restricting to privileged raiyats and persons under disability. Inaddition, it also restricted use of agricultural land for non-agricultural purpose.

Mostly tenants were landless,marginal farmers,sub-marginal farmers, and small farmers. In reality most of them belonged to class of agricultural labourers fromthe Scheduled Castes and Scheduled Tribes.

Tenants were constrained by resources to invest on land improvement and mobilizing inputs like seeds,fertilizers, and pesticides forcultivation.Credit institutions were unable to lend money to tenants who had no permanent, heritable or transferable rights over the land.Usually, mortgages were required for making long term,medium term, and short-term loans.Vesting of proprietary rights on actual tillers of land was necessary to enable them to mortgage their land. Land reforms by way of abolition of estates and vesting of transferable,heritable, and permanent rights assisted the actualcultivators to obtain ownership as understood in ordinary parlance and thus they could seek credit facilities. In this connection, it is relevant to mention that traditional money lenders in the rural areas insisted upon mortgages. Even after cooperative credit institutions and commercial bankscame, they were reluctant to give long term and medium credit without mortgage and other securities. Even for short-term crop loans pledging of crops was required. Tenants who were inducted from year to year were unable to pledge standing crops without the authority of the owners of land.

From the above it would be clear that without tenurial rights farmers could not take long or medium-term loans from the lending institutions to have land improvement measures and seek irrigation facilities like dug wells and lift irrigation. Many farmers could startsecond crop and even third crop with dug-well irrigation.Pump set on dug-well could even improve irrigation and cultivation.

Lift irrigation from the river, tank, canal or nullah required investment. Without ownership right a farmer could not give any security to the lender to raise capital.

Most of the agricultural holdings were fragmented and distributed at various locations.Absence of land consolidation and prevention of fragmentation also contributed to the constraint in agricultural productivity. Management of agriculture over land located and distributed at separate places even if they werein the same village was difficult, irksome andcumbersome. For generations, farmers struggled with fragmented land holding system.Providing irrigation, unless there was flow irrigation like major or medium irrigation facilities for the whole area,was not feasibleeconomically. It was unviable even with ownership as each fragment required water.

Law relating to consolidation of land and prevention of fragmentation was introduced in the seventies of the last century. A small farmer or marginal farmer having only two or one hectare of land in various locationswas undoubtedly burdened indoing cultivation in an efficient manner.Mechanized cultivation like use of power tiller etc. could not be easily adopted with fragmented holdings.

In Punjab land consolidation law was introduced quite early. Flow irrigation and tenurial rights added with consolidated holding gave the State head-start in agriculture production.No wonder the State became granary of the nation.

Land consolidation law came into picture in Odisha in the seventies. Implementation of the scheme under the law in the irrigated land in the ayacut of the multipurpose Hirakud dam project and in the delta irrigation project gave great encouragement to farmers to accept the consolidation law with enthusiasm while in the non-irrigated land it was resisted.

In the district of Sambalpur,Cuttack, and Puri districts the consolidation of land in the irrigated land assisted the farmers to become more productive.

In the land tenurial history of Odisha during the period when Shri Nabakrishna Choudhury was the Chief Minister of Odisha, the Estate Abolition Act was passed bringing an end to the Zamindary system, OTP and OTR Act were passed during that period and the major land reform law, OLR Act was enacted subsequently. Shri Sadashiv Tripathy as the revenue minister had played a significant rolein piloting the OLR Act. However, ceiling surplus law was implemented with commitment duringthe time of Smt. Nandini Satpathy when she was the Chief Minister. Distribution of surplus land made to agricultural labourers, land less ST/SC farmers met both economic and social objective to some extent.

Land consolidation law was enacted during the time Shri Pratap Mohanty was the revenue minister.He had taken keen interest in its implementation. He along with Shri Sudhanshu Bhushan Mishra had visited Punjab to study implementation of consolidation of land.

Contribution of civil servants like Shri Bimal Mishra, Shri Sudhanshu Mohan Patnaik, Shri Ramakant Rath and Shri Ramakant Mishra and Shri S Sundararajan were immense in development of land reform measures and land consolidation matters.

Notwithstanding all these effortsland tenurial system continues to becomplex. Scope to give ownership rights to the tillers of land continues to be vast.

For success what is needed is political will and bureaucratic commitment. In addition, tenants should be conscious of their rights and assert them at the appropriate fora.

References

Constitution of India

Orissa Tenancy Act

Orissa Estate Abolition Act

Orissa Land Reforms Act

Orissa Consolidation of Holding and Prevention of Fragmentation Act

Conclusion

Indian agriculture is a strong pillar of our economy, culture & civilization. The sector employs 54.6 percent of the workforce (Census 2011) for their primary source of livelihood and contributed to 18.6% of the country's Gross Value Added (GVA) at current price during 2021-22. Over the past 75 years, India has transitioned from being food-scarce to a food-exporting nation, driven by various science-backed agricultural revolutions implemented by the government. Yet, the challenge of meeting the food demands of an expected population exceeding 1.6 billion by 2050, necessitating a minimum 4% annual growth in agriculture, looms large (ICAR, 2022).Further, agriculture is facing new and unprecedented challenges of the 21st century- sustaining food and nutrition security, adaptation and mitigation of climate change, and sustainable use of critical natural resources like land and water. The government is continuously engaged in addressing these challenges through policy changes, and implementing programs under those. However, there is a recognized need for more such long-term directions by way of reorientation of farm policies for enhancing farmers' income and ensuring nutritional security & sustainable food system.

Phases of development in agriculture & allied sectors in India: The agricultural policies from 1950-2000 can be divided into three phases. The first phase, pre-Green Revolution, focused on agrarian reforms, irrigation projects, and institutional changes to address food shortages. The second phase, marked by the Green Revolution, saw a shift to technology driven farming, enhancing productivity and ensuring self-sufficiency in food grains. However, agrarian reforms took a back seat during this period. The third phase, post-Green Revolution until the formulation of the first National Agriculture Policy in 2000, witnessed dynamic agriculture with a focus on modernization and capital-intensive practices. Economic reforms in 1991 indirectly impacted agriculture, leading to increased subsidies and support. The National Agriculture Policy 2000 aimed at unlocking the sector's growth potential, improving rural infrastructure, promoting agribusiness, and ensuring equitable growth. The farm bills passed in 2020 aimed to liberalize trade and contract farming had to be repealed in 2021 due to wide spread farmers' protests. However, the existing policies continued to focus on modernizing small farms, attracting youth to agriculture, and fostering cooperative methods. India's vision for agriculture looks for inclusive growth and sustainability, with efforts towards organic farming, water conservation, and technological integration for climate resilience. Despite challenges, agriculture remains crucial for India's economy and food security.

Current status of Agriculture & Allied Sectors in India & Eastern States: The present status of agriculture and allied sectors in India in general, eastern India in particular has been presented in the following paragraph.

***Production of major crops*:** Over the last 75 years since independence, India has transitioned from relying on food aid to achieving self-sufficiency in food production. Food grain production has increased six-fold from 1951 to 2021. Notably, major states like Uttar Pradesh, Madhya Pradesh, and Punjab contribute significantly to this growth. India is the world's second-largest rice producer, West Bengal, Uttar Pradesh, and Punjab being the key contributors. Similarly, wheat production has seen a substantial increase in Uttar Pradesh, Madhya Pradesh, and Punjab leading to increased production. Pulses production has surged in recent years, Madhya Pradesh, Maharashtra, Haryana and Rajasthan being the major contributors. Additionally, oilseeds production has seen significant growth, particularly in states like Rajasthan, Madhya Pradesh, and Gujarat. Sugarcane production has increased substantially, led by states like Uttar Pradesh, Maharashtra, and Karnataka. Cotton production has is on rise, with Gujarat, Maharashtra, and Telangana being prominent producers. The production of horticultural crops has surpassed food grains production; states like Uttar Pradesh, Madhya Pradesh, and West Bengal have led the way. India is a major producer of fruits, with Andhra Pradesh, Maharashtra, and Uttar Pradesh being the key contributors. Cashew nut production is notable in states like Maharashtra, Andhra Pradesh and Odisha. Coconut production is significant in states like Karnataka, Tamil Nadu, and Kerala. Vegetable production is prominent in states like Uttar Pradesh, West Bengal, and Madhya Pradesh. Potato production is significant in Uttar Pradesh and West Bengal, while onion production is notable in Maharashtra and Madhya Pradesh. Tuber crops like tapioca, sweet potato, and elephant foot yam also contribute to India's agricultural output. Mushroom production has increased significantly, particularly in states like Bihar and Maharashtra. Flower production, including both cut and loose flowers, is substantial. Overall, India's agricultural landscape has witnessed remarkable growth and diversification over the years, contributing significantly to food security and economic development

***Crop bio fortification varieties*:** Besides food security, India's major initiative has been to go for food crops with better nutritional values through breeding methods. The efforts helped to develop about 85 bio-fortified varieties of different crops by 2021. These bio-fortified crop varieties are rich in protein, zinc, iron, vitamin A, vitamin C, lysine and tryptophan, 1.5 to 3.0 times more nutritious than the traditional varieties.

***Nano Liquid Fertilizers*:** The Indian Farmers and Fertilizer Cooperative (IFFCO) has developed the liquid formulations of nano Urea and nano DAP to replace the conventional forms of these fertilizers at their Nano Biotechnology Research Centre (NBRC) in Kalol, Gujarat. The size of one IFFCO Nano Urea particle is about 30 nanometers (1 nm is one-billionth of a meter), and when compared to the conventional/ granular urea, it has about 10,000 times more surface area to volume size while the size of IFFCO nano DAP is less than 100 nm. These liquid fertilizers are sprayed directly on the leaves, making systemic absorption easy for the plant. Two applications in a dose of 500 ml/ acre are ideal for the crop and 500 milliliters can substitute a full bag (45 Kg) of urea (Tomar *et al.*, 2023).

***Processing of agriculture produce*:** India's potential for processed fruits and vegetables remains largely untapped, only 2.2% of total production is commercially processed. The current processing capacity can handle just 3 to 4% of the total yield. Processed items include fruit pulps, juices, canned fruits, jams, pickles, and dehydrated vegetables. New techniques like high voltage pulse and microwave processing are gaining traction alongside traditional methods. Despite being a net exporter of agricultural products, India imports a significant amount of fresh and processed fruits and vegetables, indicating room for growth in domestic processing capabilities.

***Livestock and Poultry*:** Livestock contributes significantly to India's agricultural sector, with a GVA share of 28.36% and 5.21% of the total GVA in 2019-20. The country's livestock population stands at 536.76 million, with cattle and buffaloes comprising 36% and 20%, respectively. Efforts to improve indigenous breeds have increased their numbers. India achieved a milestone by cloning the world's first buffalo in 2009, enabling sex selection for desired offspring. Avishan sheep, developed for higher mutton production, produce 2-4 lambs per lambing. Poultry farming, which began as a backyard endeavor, has evolved into a thriving agribusiness with an annual turnover exceeding Rs 30,000 crore. India ranks third globally in egg production with an annual growth rate of 8-10%. Despite this, the per capita availability of eggs is below recommended levels. Poultry contributes 50% of the country's meat production. India is the world's largest milk producer, with a total production of

210 million tons in 2020-21. Uttar Pradesh leads in milk production, followed by Rajasthan, Madhya Pradesh, Gujarat, and Andhra Pradesh. Major eastern states also contribute significantly to milk production. Goat milk contributes 3% to the total milk production. Overall, the livestock sector plays a crucial role in India's agricultural landscape, providing livelihoods, nutrition, and economic opportunities.

***Fisheries and Aquaculture*:** India is the world's second-largest producer of aquaculture and fisheries (both marine and inland) sector's contribution to the agricultural GDP is 7.28%. It supports over 25 million people for their livelihoods. Fish production has grown from 0.75 million tons in 1950-51 to 14.164 million tons in 2019-20, with inland fisheries contributing 73.7% and marine fisheries 26.3%. Andhra Pradesh (29.49%), West Bengal(12.59%) , and Odisha(5.77%) are the top three fish-producing states, contributing 47.85% collectively. Notable advancements made include the development of 'Jayanti Rohu' with 17% higher growth, successful species diversification in freshwater aquaculture, and the adoption of technologies like marine cage-culture and composite fish culture. Marine product exports reached 1.29 million tons worth Rs 46,662.85 crore in 2019-20. Bio-floc technology has also opened avenues for entrepreneurship. Current focus areas include intensive aquaculture, carp polyculture, freshwater prawn culture, running water fish culture, and riverine fisheries development.

***Natural Resource Management*:** Land degradation, exacerbated by climate fluctuations and human activities poses a significant threat to agricultural progress in India. Approximately 37% of the country's geographical area (around 120 million hectares) suffers from various forms of land degradation, stripping of vegetation and topsoil due to mining is a serious concern. Water logging, salinity, and groundwater depletion due to poor water resource management has been impacting agricultural sustainability. Agricultural practices also contribute to climate change with methane and nitrous oxide emissions. Emission of methane is about 3.3 million tons annually in India. To mitigate these challenges, suitable technologies like Info Crop for climate impact assessment and adaptation strategies have been developed. Rainfed agriculture, practiced in 51% of India's net sown area faces substantial risks but contributes significantly to food production. Dry land technologies, integrated into national programs, aim to improve resilience through practices like rainwater harvesting and diversified cropping systems. Integrated Farming Systems (IFS) and Agroforestry models have been introduced to augment farmer income and promote sustainable land use. Organic farming has gained considerable traction through government schemes and certification systems. Despite progress, Indian agriculture grapples with land degradation, erratic rainfall, and soil health issues. Recognising need for conserving natural resources and enhancing soil fertility through inventories and mapping, scientists are engaged to develop innovative solutions including new cropping systems, integrated resource management, and participatory watershed management to ensure sustainable agricultural development.

***Agricultural engineering*:** Importance of farm mechanisation in terms of saving time & labour, reducing costs, and increasing efficiency in agricultural production and processing has been well realised. Improved manual tools and animal-drawn equipment though are ideal for small farms; availability of high-cost machinery through custom hiring is gaining popularity in rural areas. Steady increase in sale of tractors, power tillers, combine harvesters, water-saving devices etc. is the testimony to such. All of these have transformed agriculture in terms of creating new agri-businesses opportunities. Moving forward, agricultural engineering aims to advance precision farming through unmanned operations and autonomous decision support systems, addressing the challenges across the agricultural value chain.

***Agricultural Education*:** Agriculture education commenced during 1901-1905 with 6 agricultural colleges in India. Imperial Council of Agricultural Research was established on 16 July 1929, following the advice of the Royal Commission on Agriculture. By 1948, India had only 17 agricultural colleges, with 12 established since 1940. However, postgraduate research training facilities were limited, accommodating only 166 students. The Indian Agricultural Research Institute (IARI), established in 1905 at Pusa (Bihar) and later relocated to New Delhi in 1936, emerged as the nation's premier institute for agricultural research, education, and extension. Recognized as a Deemed University in 1956 by the UGC, IARI initiated masters and Ph.D. programs in various agricultural disciplines. The first veterinary college was founded in Mathura in 1947, and agricultural engineering courses leading to bachelor degrees began in 1942 at the Allahabad Agriculture Institute. The first Fisheries College was established in Mangalore in 1969. Post-independence, state Agricultural universities were established following the Land Grant pattern of the USA, recommended by the first joint Indo-American Team in 1955, aiming to upgrade agriculture and veterinary education. Today, with 74 Agricultural Universities (AUs) including State Agricultural Universities (SAUs), Deemed Universities (DUs), and Central Agricultural Universities (CAUs), India boasts one of the largest agricultural research, education, and extension systems globally. Under the New Education Policy-2020 (NEP-2020), a national-level committee has been formed by the ICAR to develop an implementation strategy, proposing significant changes in agriculture education. These changes include transforming Agricultural Universities/

Colleges into large multidisciplinary institutions, clusters, and Knowledge Hubs accommodating 3,000 or more students. The multidisciplinary approach will integrate academic programs in basic sciences, social sciences, and allied agricultural disciplines. By 2030, single-stream universities under the ICAR-AU system are expected to get transformed into multidisciplinary institutions while maintaining a focus on teaching, research, and extension in agriculture. The successful implementation of this policy will need full support from State Governments as well as amendments in the Acts and Statutes of agriculture universities in alignment with NEP-2020.

Agricultural Extension: In India, the agricultural extension system has evolved through various phases. Initially, it supported farmers in adopting intensive cultivation methods and high-yielding crop varieties for food security. Over time, programs like community development initiatives, intensive area development, Technology Transfer and Training (T&V) programs, lab-to-land initiatives, and oilseeds and pulses promotion schemes have played crucial roles. Formal extension activities were divided into mass extension by state agricultural departments and frontline extension under bodies like the Indian Council of Agricultural Research (ICAR). Currently, India's agricultural extension system comprises mass extension for dissemination of established technologies through state machinery and frontline extension involving testing and demonstration of new technologies via KVKs established by ICAR.

Agricultural Research: Established in 1929 as the Imperial Council of Agricultural Research and now known as the Indian Council of Agricultural Research (ICAR) serves as the central body overseeing agricultural research and education in India, including animal sciences and fisheries. With 113 institutions under its umbrella, including research institutes, national bureaus, project directorates, agricultural technology application research institutes, and national research centers, along with 82 coordinated research projects and networks and 74 agricultural universities, ICAR plays a pivotal role in country's agriculture development.731 Krishi Vigyan Kendra are functioning in the country. India boasts one of the largest agricultural research workforces globally, with around 30,000 scientists and over 100,000 technical and supporting personnel. Collaborations with international entities like CIMMYT and IRRI have been instrumental in India's journey towards food security, particularly through initiatives like the Green Revolution, which saw the exchange of dwarf wheat and rice varieties. Through partnerships with CGIAR and other international agricultural research centers, ICAR has expanded its efforts to develop new crop varieties, improve animal breeds, and manage natural resources. These collaborations have led to advancements such as the development of new livestock breeds, vaccines, diagnostics, and quality standards, contributing to enhanced health, nutrition, and hygiene standards in the livestock and fisheries sectors.

Development of Fisheries and Aquaculture in Eastern India: Fisheries and aquaculture are crucial to the food security and economy of eastern India, which includes West Bengal, Bihar, Jharkhand, and Odisha, home to 31.66 crore people. These states benefit from rich aquatic resources such as rivers, reservoirs, ponds, and wetlands. Historically, this region has been pivotal in the development of fisheries technologies. Despite significant inland fish production, growth in the eastern states has lagged behind national trends due to challenges like limited scope for intensification in homestead ponds. Odisha and West Bengal are leading fish producers with considerable contributions from both inland and marine sources, while Bihar and Jharkhand, though facing challenges like extreme weather, have also made notable progress in aquaculture development. In 2022, Eastern India's marine fisheries saw West Bengal and Odisha contribute 9% of India's total landings, with West Bengal leading at 1.90 lakh tonnes (58.8%) and Odisha at 41.2%. Both states' marine resources are similar, with pelagic resources dominating. West Bengal's major landings were anchovies and penaeid prawns, while Odisha's included penaeid prawns and croakers. Mechanized fishing dominates both states, with the third quarter being the peak season. West Bengal's marine fish landings averaged 2.085 lakh tonnes annually from 2013-22, valued at Rs 3153 crores in 2022, while Odisha's average was 1.3 lakh tonnes, valued at Rs 2521 crores. Both states show healthy fish stock assessments, though some species like the tiger-tooth croaker are overfished. Inland fisheries in these states, including riverine, reservoir, and wetland fisheries, face challenges such as habitat degradation and insufficient infrastructure but show potential for growth through improved management and technological interventions. Aquaculture development in India has evolved significantly since the mid-20th century, driven by efforts from key research institutions such as the Central Inland Fisheries Research Institute (CIFRI) and the Central Institute of Freshwater Aquaculture (CIFA). Early advancements included the bundh breeding technique for fish seed production, which was later complemented by induced breeding methods starting in the 1950s. Freshwater aquaculture saw considerable growth in Odisha, West Bengal, Bihar, and Jharkhand, with Odisha increasing its fish production from 2.24 lakh tonnes in 2010 to 6.90 lakh tonnes in 2022-23. Similarly, West Bengal and Bihar have developed extensive fish seed production networks, with West Bengal emerging as a leader in seed production and Bihar making strides in scientific fish farming. Jharkhand has also advanced in cage farming and fish seed production. These developments have been supported by improvements in breeding techniques, including the successful use of hormonal induction for controlled fish breeding. Enclosure aquaculture, particularly cage and pen culture, has significantly boosted fish production in large water bodies, with ICAR-CIFRI leading advancements in technology and implementation across states such as Assam, Bihar, Chhattisgarh, Jharkhand, and Odisha. These methods enhance stock through systematic design, species selection, and ecosystem impact studies. In brackish water aquaculture, ICAR-CIBA has pioneered diverse finfish and crustacean cultures, particularly in West Bengal and Odisha, where shrimp farming dominates. Despite challenges such as disease outbreaks and low-density farming, advancements in feed production and health management have improved fish farming efficiency. The Blue Revolution has dramatically increased fish production in India, supported by policy, infrastructure, and scientific research, with ongoing efforts to address future challenges like environmental impacts and resource constraints.

Livestock - a lifeline for small and marginal farmers of eastern India: The livestock sector is vital for rural household development in India, especially for the underprivileged, offering more equitable opportunities than land ownership. Over the past decade, its growth has surpassed that of the crop sector, driven by increasing demand, population growth, and urbanization. In 2020-21, livestock contributed significantly to the agricultural sector's GDP, accounting for 30.86% of the Gross Value Addition (GVA). States like Bihar and Odisha have demonstrated remarkable growth, with livestock contributing 38% and 20% respectively to their agricultural output. However, the sector faces challenges such as poor infrastructure, low productivity, disease outbreaks, and feed availability. The Odisha government's "Odisha Bovine Breeding Policy, 2015" and the National Livestock Mission (NLM) launched in 2014-15 by the Government of India aim to address these issues through holistic development. Rising per capita income, urbanization, and global demand for livestock products present significant opportunities for growth and poverty alleviation. Empowering small-scale producers, particularly women, is essential to harness these opportunities. Creating an enabling environment through targeted policies and infrastructure development is crucial for realizing the sector's full potential and improving the livelihoods of millions.

Concepts in Cattle Breeding for Sustainable Improvement: The depletion of native breeds is driven by crossbreeding with exotic breeds, reduced economic viability, small herd sizes, and agricultural mechanization. Genetic diversity Conservation is crucial for scientific research, ecosystem balance, cultural heritage, ethical reasons, and future energy sources. Native breeds, untainted by exotic semen, are ideal for producing A2 milk, which is safer and healthier. Odisha's four cattle breeds (Binjharpuri, Ghumusari, Khariar, and

Motu) and three buffalo breeds (Chilika, Kalahandi, and Manda) excel in A2 milk production with high disease resistance. Chilika buffaloes, in particular, produce milk with a favorable n-3 to n-6 PUFA ratio, beneficial to human health. Increasing A2 milk production requires extensive rearing of native breeds. It requires supportive conservation policies and increased awareness among consumers, that would help reducing reliance on crossbreeding and artificial insemination.

Promoting Poultry for Eastern India: Emphasis on Backyard Poultry Production in Odisha: Poultry production has seen significant growth and expansion across West Bengal, Assam, Bihar, Jharkhand, Chhattisgarh, and Odisha, largely due to government support and initiatives. In West Bengal, production tripled from 2000 to 2020, reaching 1.2 million tons, supported by state schemes. Similarly, Assam, Bihar, Jharkhand, and Chhattisgarh witnessed production increases of two to three times during the same period. Government of Odisha launched "Odisha Poultry Policy 2015" to foster poultry development with ambitious production targets. APICOL, a state owned corporation provides support and guidance to poultry enterprises, including capital investment subsidies. Despite progress, per capita egg and meat availability still lag behind national averages, presenting opportunities for improvement. Indigenous breeds like Hansli and native ducks are vital to Odisha's rural economy alongside the growing popularity of quail farming. Backyard poultry farming, a key source of supplementary income for rural households is being encouraged with emphasis on improved germplasm and management practices. However, the introduction of crossbred poultry raises concerns about native breed preservation. Efforts are underway to address challenges and enhance the role of poultry in nutritional security, livelihood support, and poverty reduction, particularly on women empowerment. Extension support is crucial for addressing issues like germplasm availability, housing management, and health care for backyard poultry farming. Despite slower growth compared to other regions, Odisha's poultry sector is steadily progressing, reflecting the broader trend of poultry development in the Eastern region of India.

Role of Women in Agriculture and Strategies for Gender Mainstreaming: In rural India, women face numerous challenges including limited access to land, education, credit, and technology. Despite their higher involvement in agricultural work, they often receive lower remuneration than men. They often face challenges in balancing multiple roles with little time left for own selves. Poor nutrition in early life hampers their health and earning potential, exacerbating the cycle of poverty. For improving agricultural productivity and achieving Sustainable Development Goals, particularly in combating violence, inequality, and poverty, it is imperative to find solutions to these challenges. Hence, gender equality has been viewed as fundamental to SDGs, especially in tackling hunger and extreme poverty. The primary source of income for women in India (80%) is agriculture, of which 33 per cent as agricultural labor force and rest 48 per cent as self-employed farmers. However, their productivity is hindered by time constraints, lack of resources, and unequal access to services. Key issues for women in agriculture include their triple role, involvement in labor-intensive tasks, limited access to resources, and unequal recognition and remuneration. To ensure economically and ecologically sustainable agriculture, both men and women must actively participate in modernizing the sector. This requires upgrading women's skills, developing gender-friendly technologies, promoting women's groups, and ensuring equal access to resources. Addressing these challenges necessitates gender equity in planning and policy formulation, as well as fostering social and cultural changes to recognize and empower women farmers through their meaningful participation in decision-making and program implementation. Ultimately, India's agricultural transformation hinges on educated, skilled, and visionary women farmers who can lead the sector into a dynamic and innovative future.

Tools and Approaches in Agricultural Extension in Eastern India: In pursuit of sustainable agricultural development, a nation's success heavily hinges on the efficacy of its agricultural extension strategies, methodologies, service delivery, and processes. Despite efforts made by State/ Central Govt, a noticeable gap persists in agronomic practices and socio-economic outcomes between North Western and Eastern India. Bridging of these gaps requires an embrace of extension tools and methodologies rooted in research and tailored to address the unique development needs of each region. In Eastern India, the extension system demonstrates a pluralistic approach, employing a diverse array of extension tools and methods. Key extensions include the utilization of front-line extension services offered by institutions such as ICAR (Indian Council of Agricultural Research) and SAUs (State Agricultural Universities), as well as KVKs (Krishi Vigyan Kendras), ATICs (Agricultural Technology Information Centers), and ABI (Agri Business Incubation) under various institutions. NMAET (National Mission on Agricultural Extension and Technology) under the Ministry of Agriculture, Government of India and State departments focusing on agriculture, horticulture, animal husbandry, and fisheries are also playing important roles. Organizations like MANAGE (National Institute of Agricultural Extension Management), NIPHM (National Institute of Plant Health Management), IIPM (Indian Institute of Plantation Management), EEI (Extension Education Institute), and SAMETI (State Agricultural Management & Extension Training Institute) contribute significantly to crafting extension policies, implementing programs, and enhancing the skills of extension professionals. Recent extension reforms have introduced public participatory extension models like ATMA (Agricultural Technology Management Agency), franchised convergence extensions such as the IARI (Indian Agricultural Research Institute)-Post Office Linkage Extension model, and collaborations with CGIAR (Consultative Group for International Agricultural Research) institutes like CSISA (Cereal Systems Initiative for South Asia). Extension approaches targeting specific development objectives, such as the Complex Driven Rainfed Plan, TSP (Tribal Sub-Plan), SCSP (Scheduled Caste Sub Plan), State Plan-RKVY (Rastriya Krishi Vikas Yojana), regional plans like BGREI (Bringing Green Revolution to Eastern India), Aspirational District Plan, NFSM (National Food Security Mission), and Mission for Integrated Development of Horticulture MIDH-National Horticulture Mission (NHM), are also being implemented by the field formations of concerned govt. departments. Various public-private partnership extension models such as government-regulated private extension, government-franchised private extension, government-franchised grassroots-led extension, government-supported Agri-graduates ACABC (Agri-Clinics and Agri-Business Centers) Extension, and para-professional-led extension services have also emerged over time. Additionally, there are privatized extension models such as Input Agency Extension, Consultancy Service Sector Extension, Financial Agencies Extension, Market-led Extension, Contract Farming Extension, and NGO-led Extension. In the digital era, cyber extension methods like Farmer Portals, Apps-computer software, Television channels, AIR (All India Radio), and m-Kissan-Portal of Government of India for Farmer Centric Mobile Based Services are helping disseminate crucial messages to a broader audience. Gender-sensitive extension models like the Women Para Extension Worker Model and Mission Shakti are making a positive impact in Eastern India. Cooperative and FPO (Farmers Producer Organization) based group-led extensions, along with agriculture fairs, exhibitions, awareness campaigns, diagnostic visits, surveillance, mass extension approaches, demonstrations, field days, training sessions, and meetings, are among the commonly used group approaches in Eastern India. These innovative methods are reshaping the role of extension services, serving as vital links between researchers and farmers while fostering collaborative and participatory research extension strategies. There is a multilevel partnership, convergence, franchising, community involvement, and ICT (Information and Communication Technology) – mediated pluralistic extension system already in place. Extension approaches with more comprehensive understanding of current extension dynamics, considering the diversity and varied needs of different farming systems is called for.

Challenges in Agriculture Sector and Role of Financial Institutions in Eastern India: There is an wide variance in per capita NSDP and agricultural credit deployment among the eastern states like Bihar, Jharkhand, Odisha, and West Bengal. NABARD supports agriculture through schemes like RIDF and Commercial Banks as well as RRBs provide credits to the agriculture sector under priority sector lending, still enough room left for improvement. RRBs and Cooperative Banks have been playing significant role in rural areas and Small Finance Banks have been catering to the underserved areas. MFIs provide collateral-free loans but are limited in these states. Credit penetration and branch coverage need enhancement, efforts by the government and RBI for financial inclusion in rural areas is imperative. Collaborative efforts involving SLBC, NABARD, and state agencies are crucial for maximizing agricultural potential in the region.

National Agro-forestry Policy 2014 in India: The National Agroforestry Policy of India, launched in February 2014, aims to enhance agricultural livelihoods by integrating agriculture, livestock farming, and forestry on the same land. This policy, introduced during the World Congress on Agroforestry in Delhi, focuses on increasing agricultural and animal productivity, expanding tree cover, and mitigating climate change. Agroforestry, which combines these practices, offers numerous benefits including higher productivity and various social, economic, and environmental advantages. Proper farm management is crucial for realizing these benefits. India became the first country to adopt an inter-sectoral policy on agroforestry, aiming to remove constraints and ensure other policies support agroforestry. The policy bridges forestry, agriculture, water, and environmental sectors, supporting India's goal of 33 percent tree cover while also enhancing food security, nutrition, and providing resources like fodder, fuel wood, and timber.Agroforestry is seen as a vital tool for building resilience against climate change and natural disasters, as well as promoting rural employment and development through agroforestry-based economic opportunities.To achieve the deliverables of the National Agroforestry Policy, several key strategies are outlined. These include integrating agroforestry into agricultural policies, creating a dedicated fund to support coordinated interventions, and simplifying state regulations related to forestry and land use. States should identify and exempt 20 common tree species from regulatory constraints, ensure secure land tenure, and promote public-private partnerships for agroforestry on non-forest lands. Providing quality planting materials through accredited nurseries, collecting data on agroforestry produce, and encouraging research on indigenous species are essential. The policy advocates for upgrading the National Research Centre for Agroforestry, establishing effective extension mechanisms involving universities and KVKs, and incorporating agroforestry into educational curricula. Additionally, it calls for developing marketing infrastructure, exploring contract farming, and providing financial incentives and insurance for agroforestry farmers. The policy also aims to quantify carbon sequestration benefits and promote industry involvement in value chain development and technology.

Horticultural policies and development in India: India's horticulture sector is vital for livelihoods and food security, especially in the face of climate challenges. Emphasizing research on climate-resilient varieties and sustainable practices is essential. Promoting organic and niche products can help access premium markets. Cultivating exotic fruits and vegetables presents new income opportunities for farmers, showcasing the sector's potential for growth and innovation.

The development of horticulture in India has evolved through distinct phases, each marked by significant shifts in policy, technology, and market dynamics. Initially focused on subsistence farming with limited commercial cultivation pre-independence, the sector saw gradual advancements during the Green Revolution era, though primarily overshadowed by efforts to achieve food security. From the 1980s onwards, initiatives like the National Horticulture Mission and subsequent programs spurred substantial growth, introducing advanced technologies, improving post-harvest management, and expanding market opportunities. Liberalization from the 1990s further catalysed sectoral growth, integrating private sector participation and enhancing export capabilities. Recent years have seen comprehensive development under the Mission for Integrated Development of Horticulture, emphasizing quality improvement, digital tools, sustainable practices, and value-added products. These phases collectively illustrate horticulture's transformation from traditional practices to a technology-driven, market-oriented sector crucial for food security, income generation, and economic growth in India.

The horticulture sector, despite its transformative potential, faces challenges such as inadequate post-harvest infrastructure, absence of Minimum Support Price (MSP) for perishables, and insufficient transportation and storage facilities. Addressing these issues through strategic interventions like research and development, resource optimization, and value addition can significantly enhance the sector's profitability and market access. Research in India, led by ICAR and other institutions, has advanced crop management practices, developed disease-resistant varieties, and introduced innovations in propagation, irrigation, and post-harvest technologies, boosting productivity and quality. Government initiatives like the Mission for Integrated Development of Horticulture (MIDH) aim to foster holistic growth, supported by policies promoting pollinator-friendly practices and diversification strategies. Collaboration among stakeholders and integration of modern technology are crucial for unlocking the sector's potential, enhancing sustainability, and ensuring resilience against climate variability.

Land Tenure and Agricultural Productivity: Land tenure rules dictate how property rights to land are managed in a society, impacting agricultural growth. Historically, India's land tenure was influenced by intermediaries like the Zamindary system, abolished post-independence. The Orissa Estate Abolition Act of 1953 dismantled intermediaries, yet ownership didn't automatically transfer to tillers. Subsequent reforms, including the Orissa Tenancy Reforms Act and Land Reforms Act of 1960, aimed at redistributing land to marginalized farmers and limiting non-agricultural use. However, tenants, often landless or small holders continue to face difficulties in investment, access to credit, and land consolidation. Despite efforts, complexities persist in the tenurial system, highlighting the need for further reforms to empower land tillers.

References

A, Balogh KD, Garcia AD. Antimicrobials used in backyard and commercial poultry and swine farms in the Philippines: a qualitative pilot study. Frontier in Veterinary Science. 2020; 7:329.

Agriculture Statistics at a Glance,2020, & 2021 Government of India, Ministry of Agriculture and farmers Welfare, Department of Agriculture and Farmers Welfare, Directorate of Economics and Statistics

Ahlers C, Alders R, Bagnol B, Cambaza AB, Harun M, Mgomezulu R, Msami H, Pym B,

Area and Production of Horticulture Crops for 2018-19 (Third Advance Estimates) and 2019-20 (Third Advance Estimates), Department Agriculture & Farmers' welfare, Government of India

Annual Report, 2021-22, Government of India, Ministries of Fisheries, Animal Husbandry and Dairying

Annual Report, 2020-21& 2021-22 Government of India, Ministry of Agriculture and farmers Welfare, Krishi Bhawan, New Delhi

Basavaraja, N. 2007. Freshwater fish seed resources in India, pp. 267–327. In: M.G. Bondad-Reantaso (Ed.). Assessment of freshwater fish seed resources for sustainable aquaculture. FAO Fisheries Technical Paper. No. 501. Rome, FAO. 2007. 628p.

Basic Animal Husbandry Statistics, 2019, Government of India, Ministry of Fisheries, Animal Husbandry and Dairying, Krushi Bhawan, New Delhi

Biofortified Varieties: Sustainable Way to Alleviate Malnutrition,3rd edition, ICAR, New Delhi,2020

CMFRI, FRAEED (2023) Marine Fish Landings in India - 2022. Technical Report. ICAR-Central Marine Fisheries Research Institute, Kochi.

Das, L., Mishra, S.K. and Rath, N.C. 2006. Decent work and empowerment of women in agriculture. Kurukshetra, 54(7):21-28.

DoF, 2022. Hand Book of Fisheries Statistics, Department of Fisheries, Ministry of Fisheries, Animal Husbandry & Dairying, Govt. of India.

Economic Survey, 2021. Agriculture & Food management, Ministry of Finance, Government of India.2:230-260. Available athttps://www.indiabudget.gov.in/economicsurvey/.

Hand book of Statistics on Indian States, 2020-21, Reserve bank of India,

Hand book of Statistics on Indian Economy, 2021-22, Reserve bank of India, Handbook of Statistics on Indian Economy 2021-22 – https://rbi.org.in, https://dbie.rbi.org.in

Hand book on Fisheries Statistics, 2020, Department of Fisheries, Ministries of Fisheries, Animal Husbandry and Dairying, , Government of India, New Delhi

H. Pathak, J.P. Mishra, T. Mohapatra (2022), Indian Agriculture after Independence, Indian Council of Agricultural research, New Delhi

https://www.agric.wa.gov.au/irrigated-crops/nutritional-aspects-quinoa

Kumar Ujjwal, Singh, D, K,, Bhatt, B, P,, Sarkar Bikash, Koley, T. K. and Gupta Santosh (2017). Mar¬ket Led Agricultural Extension-Concept and Practices. Training Manual ICAR Research Complex for Eastern Region, Patna -800 014

National Agroforestry Policy, Government of India, department of Agriculture & Cooperation. Ministry of Agriculture, New Delhi

Orissa Tenancy Act

Orissa Estate Abolition Act

Orissa Land Reforms Act

Singh, Krishna M., Meena, Mohar Singh and Swanson, Burton E. (2013). Role of State Agricultural Universities and Directorates of Extension Education in agricultural extension in India. Available at SSRN: https://ssrn.com/abstract=2315381 or http://dx.doi.org/10.2139/ssrn.2315381